FRACTURE MECHANICS – WORKED EXAMPLES

Fracture Mechanics
Worked Examples

J.F. Knott FRS, FEng, B.Met, ScD, FIM, FWeldI
School of Metallurgy and Materials
The University of Birmingham
and
P.A. Withey BSc, PhD, CPhys, MInstP
Rolls-Royce plc
Bristol

CRC Press
Taylor & Francis Group
Boca Raton London New York

CRC Press is an imprint of the
Taylor & Francis Group, an **informa** business

CRC Press
Taylor & Francis Group
6000 Broken Sound Parkway NW, Suite 300
Boca Raton, FL 33487-2742

Visit the Taylor & Francis Web site at
http://www.taylorandfrancis.com

and the CRC Press Web site at
http://www.crcpress.com

Contents

List of Worked Examples

Acknowledgements

The authors would like to thank Dr. D. Elliott of BAeSEMA for his work on the first edition of *Worked Examples in Fracture Mechanics* and his help and encouragement on this second edition.

The authors would also wish to acknowledge the assistance given by the following individuals:

Prof. J.T. Barnby,   Dr. R.S. Brooks,   Dr. R.S. Pilkington,   Dr. C.A. Hippsley, Mr. E.F. Walker, Mr. S.E. Webster

and others who, knowingly or unknowingly, have contributed to the compilation of this book.

Foreword

Fracture Mechanics has developed rapidly from a largely theoretical base to one of important practical consequence in the field of component design, service inspection and safe materials usage.

The main purpose of this publication is to introduce the reader to the concepts of fracture mechanics by a series of WORKED EXAMPLES, to illustrate how fracture toughness values are derived and to show how this information can be utilised as a means of avoiding failures in service.

The book is not intended for experts in fracture mechanics, but for those who wish to learn about the subject and its application.

1. Introduction

Conventional engineering design is based on avoidance of failure by general plastic collapse. The material property specified in design codes is the flow stress: usually the yield stress or 0.2% proof stress, but occasionally, in older codes, the tensile strength. The DESIGN STRESS is then the applied stress calculated to cause collapse, divided by a SAFETY FACTOR. Typical safety factors are: 1.5 for wrought steel in applications such as pressure vessels or boilers; perhaps, 4 for steel castings in similar applications; and some 5-10 for wire ropes, supporting crane hooks or lift cages. The prime aim of the safety factor is to take account of any extra stresses imposed during erection, fabrication, or service, which may raise the applied stress to the value required to cause plastic collapse and failure.

As defined above, the safety factor does not recognise the possibility of failure by an alternative mode such as 'brittle' or 'fast' fracture. It was generally believed that the safety factor could 'safeguard' against this type of low stress fracture by the use of higher figures applied to the tensile strength. However experience has shown this not to be the case, there being a number of instances where total failure of a component or structure has occurred in the presence of a material defect or crack at stresses well below the design stress. Moreover, the higher safety factors applied for castings, as compared with wrought material, stem from fears that the castings might contain more inherent defects, which could lead to fast crack propagation at or below the applied design stress. This, in the engineering sense, is a 'brittle' failure and it is clearly necessary that a STRESS CONCENTRATOR must be present to obtain a brittle failure because the plastic strain required to operate the fracture mechanism has to be able to develop in a local region, without causing overall general collapse.

In service, the stress concentrators of importance are CRACK-LIKE DEFECTS, particularly if these are situated in regions of high background stress, such as those around fillets, keyways, nozzle openings or hatchways. Typical examples of crack-like defects include:-

> Solidification cracking in welds or castings
> Hydrogen cracking in heat-affected zones
> Lamellar tears around inclusions in rolled plate
> Cracks which have grown in a 'sub-critical' manner by fatigue or
> stress-corrosion mechanisms.

It is usually possible to detect such defects, using ultrasonic inspection or some other NDT technique and to determine the maximum size of defect in the region of interest.

1

The aim of FRACTURE MECHANICS is then to calculate whether or not a defect of given size will propagate in a catastrophic manner under service loading and thence to determine the degree of safety that the structure possesses with respect to failure by fracture. The property which measures resistance to fast crack propagation is the material's FRACTURE TOUGHNESS (measured by loading to fracture testpieces which contain sharp cracks of known lengths).

In this new edition, we have extended coverage to include the J-integral methods and some aspects of sub-critical crack growth. Details of test procedures have been revised to conform to BS 7448 (1991).

Units

The units used throughout this book are taken from the SI system of units. However, other systems of units are used within the study of fracture mechanics and a conversion table of the most commonly used units and their conversion factors is given below.

To convert from	To	Multiply by
inch	metre (m)	2.54×10^{-2}
pound force	newton (N)	4.448
kilogram force	newton (N)	9.807
kilogram force/metre^2	pascal (Pa)	9.807
pound mass	kilogram (kg)	4.536×10^{-1}
ksi	pascal (Pa)	6.895×10^{6}
ksi $\sqrt{in}$	MPa m$^{1/2}$	1.099
daNcm$^{-3/2}$	MPa m$^{1/2}$	1×10^{2}

Table 1.1

Further difficulty may arise in the interpretation of SI units. Due to the magnitude of the numbers involved the stress within a body is usually given in MPa. However MPa may be calculated in one of two ways, either from MN and m or from N and mm. This will cause no difficulty until fracture toughness is calculated. If the units used are N and mm the result will be 31.623 larger than the result using MN and m. Care has to be taken to enter the data into the equations in the correct form to obtain the correct result. MN and m are used through this book.

This change of units can cause problems in equations such as the Paris Law (equation 54)

$$da/dN = A\Delta K^m$$

in which the constants in the equation are dimensional and therefore vary with a change in units. The value of m, and the units of K and da subsequently affect the value of A, hence the values of these constants must be given with reference to the system of units used (in this book MPa m$^{1/2}$ and m respectively are used).

2. Background Theory

2.1 Griffith's Relationship

The basic theory on which fracture mechanics is founded emanates from the work of A.A. Griffith in 1920. This concerned the calculation of the fracture strength of a brittle solid (glass) which contained a sharp crack. The model, Figure 2.1, is that of a through-thickness crack of length $2a$ in an infinite body, lying normal to a uniform applied tensile stress, σ_{app}.

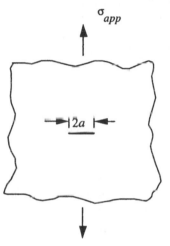

Figure 2.1

Plane-strain conditions are assumed: i.e. a condition of zero strain in the direction orthogonal to both the crack length and that of the applied stress.

An energy balance equation, assuming linear elastic behaviour, then gives for the fracture stress, σ_F:

$$\sigma_F = \left(\frac{2E\gamma}{\pi\left(1-v^2\right)a} \right)^{1/2} \qquad (1)$$

where

E is Young's Modulus
v is Poisson's Ratio
2γ is the work of fracture (γ is often taken as the surface energy)

This expression provides therefore a relationship between fracture stress and crack length if the material's work of fracture (2γ) is known.

2.2 Orowan / Irwin Relationship

The above relationship was later modified by Orowan and Irwin to take account of the occurrence of plastic flow at the crack tip before the onset of crack extension. Then, using elastic relationships for the body as a whole, which can be justified only if the size of the plastic zone is very small, the equation becomes:

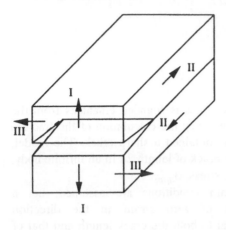

Figure 2.2

$$\sigma_F = \left(\frac{EG_{Ic}}{\pi(1-v^2)a}\right)^{1/2} \qquad (2)$$

where G_{Ic} is the material's plane strain (opening mode *I*) fracture toughness. In metals, G_{Ic} is a measure primarily of the amount of plastic work which must be done before the crack extends.

It is also possible to apply stress to the crack in plane shear (mode *II*) or antiplane shear (mode *III*), where the toughnesses are written as G_{IIc} and G_{IIIc} respectively. These are illustrated in Figure 2.2

In plane-stress deformation (very thin sheet) the $(1-v^2)$ factor is missing from the denominators of the previous equations and the material's toughness is then written as G_c.

Usually, Equation (2), or its plane-stress version, is used to calculate the maximum size of defect that can be tolerated under a given design stress. This process can however be reversed to estimate the maximum stress that can be applied to a component which contains a crack of known length.

Example 2a - Calculation of Minimum Defect Size

Rocket motor cases may be fabricated as thin walled tubes from:
(a) low alloy steel of proof stress 1200 MPa and fracture toughness 24 kJ m^{-2} measured in sheet of the appropriate thickness, or
(b) maraging steel of proof stress 1800 MPa and fracture toughness 24 kJ m^{-2}.

In a particular application, the design code specifies that the design stress is $\sigma_y/1.5$, where σ_y is the proof stress. Calculate the minimum size of defect required to give brittle fracture in service for the two materials. Comment on the result. (Young's modulus may be taken as 200 GPa in both cases).

Solution

For thin sheet, use Equation (2) in plane stress form *viz*:

$$\sigma_F = \left(\frac{EG_c}{\pi a}\right)^{1/2}$$

If fracture is to occur in service, we need to calculate the value of '*a*' when

$$\sigma_F = \frac{\sigma_y}{1.5}$$

Case (a) - Low Alloy Steel
(work in MN and m)
$\sigma_y = 1200$, $\sigma_y/1.5 = 800$, $G_c = 24$ kJ m^2

$$\therefore \; 800 = \left(\frac{200 \times 10^3 \times 24 \times 10^{-3}}{\pi a}\right)^{1/2}$$

$$64 \times 10^4 = \frac{48 \times 10^2}{\pi a}$$

$$\therefore \; a = \left(\frac{3}{4\pi}\right) \times 10^{-2} \, m$$

$$= 2.40 \text{ mm}$$

This is the half-length of a central crack. **The minimum crack length is therefore 4.80 mm.**

Case (b) - Maraging Steel
$\sigma y = 1800$, $\quad \sigma y/1.5 = 1200$, $\quad G_c = 24$ kJ m^{-2}

$$\therefore \; 1200 = \left(\frac{200 \times 10^3 \times 24 \times 10^{-3}}{\pi a}\right)^{1/2}$$

$$\therefore \; a = \left(\frac{1}{3\pi}\right) \times 10^{-2} \, m \; = 1.06 \text{mm}$$

Therefore the minimum crack length is **2.12mm**.

Comments

We see that, because the design stress is given as a fraction of the proof stress, it is necessary to increase the toughness of the maraging steel by a factor of $(1800 / 1200)^2$ to a G_c value of 54 kJm^{-2} to give the same tolerance to defects.

Example 2b - Calculation of Fracture Stress

If the fracture stress of a large sheet of maraging steel containing a central crack of length 40 mm is 480 MPa, calculate the fracture stress of a similar sheet containing a crack of length 100 mm.

<u>Solution</u>
(Work in MN and m)
From Equation 2,

$$\sigma_F = \text{const. } a^{-1/2} \text{ where } 2a = 40\text{mm}$$

Hence, $480 = \text{const. } (0.02)^{-1/2}$

$$\therefore \text{ const.} = 480(0.02)^{+1/2}$$

For $2a = 100$ mm,

$$\sigma_F = \text{const. } (0.05)^{-1/2}$$

$$\therefore \ \sigma_F = 480\left(\frac{0.02}{0.05}\right)^{1/2}$$

$$= 304 \text{ MPa}$$

2.3 Energy Release in Terms of Crack Tip Stresses

Most practical situations cannot be modelled by the simple infinite body configuration and allowance must be made for the presence of free surfaces, or combinations of applied stresses.

A commonly used method of deriving crack tip stresses is by stress analysis calculation, whilst a more direct experimental approach uses compliance techniques (see Section 2.6).

In its simplest form, the tensile stress distribution ahead of a sharp, through-thickness crack of length 2a in an infinite body (shown schematically in Figure 2.3) is given by:

$$\sigma = \frac{\sigma_{app}}{\left(1-a^2/x^2\right)^{1/2}} \tag{3}$$

which holds in the regions $-x < -a$ and $x > a$. To understand why the stress distribution is of this form see the Appendix.

If we consider Equation 3, in relation to Figure 2.3, we see that:

Near the crack tip, $x \to a$ and $\sigma \to \infty$

At large distances, $x \to \infty$, $a/x \to 0$, $\sigma \to \sigma_{app}$

If Equation 3 is now written in terms of the distance ahead of the crack tip, $r = (x-a)$, the local stress *very close* to the crack tip (i.e. $r \ll a$) becomes:

$$\sigma = \frac{K}{(2\pi r)^{1/2}} \tag{4}$$

where K is defined as the STRESS INTENSITY FACTOR and has the value $K=\sigma_{app}(\pi a)^{1/2}$ for a central crack of length $2a$ (see Table 2.1 later). K has the units of MN m$^{-3/2}$ or MPa m$^{1/2}$, in the S.I. system based on MN and M. (See Table 1.1 for conversion factors to other sets of units).

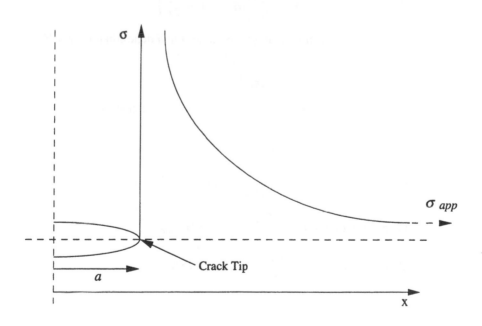

Figure 2.3

Example 2c - Derivation of Stress Intensity Factor

Show that equation 4 is a first approximation to the stress ahead of a crack, but is about 1.5% inaccurate at a distance $r = 0.02a$ ahead of the crack.

Solution

From Equation3, $\sigma = \dfrac{\sigma_{app} x}{\left(x^2 - a^2\right)^{1/2}}$

let $r = (x - a)$ and substitute $x = (r + a)$.

Then
$$\sigma = \dfrac{\sigma_{app}(r+a)}{\left[\left(r^2 + 2ar + a^2\right) - a^2\right]^{1/2}}$$

$$= \dfrac{\sigma_{app}(r+a)}{\left(r^2 + 2ar\right)^{1/2}}$$

Now, if $r \ll a$, $(r + a) \to a$ and $(r^2 + 2ar) \to 2ar$.

Hence,
$$\sigma = \dfrac{\sigma_{app} a}{(2ar)^{1/2}} = \sigma_{app}\left(\dfrac{a}{2r}\right)^{1/2}$$

To follow convention, multiply numerator and denominator by $\pi^{1/2}$

$$\therefore \sigma = \dfrac{\sigma_{app}(\pi a)^{1/2}}{(2\pi r)^{1/2}}$$

or
$$\sigma = \dfrac{K}{(2\pi r)^{1/2}} \qquad\qquad \text{(cf. Equation 4)}$$

At $r = 0.02a$, the value of x is $1.02a$

From Equation 3, $\sigma = \dfrac{\sigma_{app} 1.02a}{\left(0.0404a^2\right)^{1/2}} = 5.075\sigma_{app}$

From Equation 4, $\sigma = \dfrac{\sigma_{app}}{(0.04)^{1/2}} = 5.000\sigma_{app}$

Thus, the % error $= \left(\dfrac{0.075}{5.075}\right) \times 100 = \mathbf{1.48\%}$

In the derivation of the Griffith's relationship, Equation 1, the propagation of a crack is treated thermodynamically, by balancing the elastic energy released when a crack extends against the work required to produce two new surfaces. A particularly powerful method of calculating the change in energy, $\delta\xi$, when a crack is extended by an amount, δa, makes use of a virtual work argument - the crack tip stress field does work as it reduces to zero by moving through the displacements of the virtually extended crack.

Example 2d - The Virtual Work Theorem

Show that, if the term $d\xi/da$ per unit thickness is defined as the POTENTIAL ENERGY RELEASE RATE, G, it is possible to express G in terms of the stress intensity factor, K, through the relationship:

$$G = \frac{\alpha K^2}{E}$$

where $\alpha = 1$ in plane stress

and $\alpha = (1 - v^2)$ in plane strain.

Solution

The stress ahead of a crack of length a is given by:

$$\sigma = K \, (2\pi r)^{-1/2} \quad \text{see Equation 4}$$

The displacement within a crack virtually extended to $(a + \delta a)$ is given by:

$$u = 2\sqrt{\frac{2}{\pi}} \alpha \frac{K}{E} (\delta a - r)^{1/2} \quad \text{(from the Appendix)}$$

where r is still measured from the tip of the unextended crack.
Then the work done, *per unit thickness*, on virtual crack extension is given by:-

$$\delta\xi = \int_0^{\delta a} \sigma u \, dr = \int_0^{\delta a} \frac{K}{(2\pi r)^{1/2}} u \, dr$$

$$= \frac{2}{\pi} \alpha \frac{K^2}{E} \int_0^{\delta a} \left(\frac{\delta a - r}{r} \right)^{1/2} dr$$

Substitute $r = \delta a \, \sin^2\theta$ $r = 0 \; \theta = 0$

$$r = \delta a \quad \theta = \pi/2$$

$$dr = 2\,\delta a\,\sin\theta\,\cos\theta\,d\theta$$

$$\int_0^{\delta a}\left(\frac{\delta a - r}{r}\right)^{1/2} dr = \int_0^{\pi/2} 2\left(\frac{\cos\theta}{\sin\theta}\right)\sin\theta\cos\theta\,\delta a\,d\theta$$

$$= \int_0^{\pi/2}(1+\cos2\theta)\delta a\,d\theta$$

$$= \left[\theta + \frac{1}{2}\sin2\theta\right]_0^{\pi/2}\delta a$$

$$= \frac{\pi}{2}\delta a$$

Hence,
$$\delta\xi = \alpha\frac{K^2}{E}\delta a$$

$$\frac{\delta\xi}{\delta a} = \alpha\frac{K^2}{E}$$

In the limit $\delta a \to 0$, $\quad \dfrac{\delta\xi}{\delta a} = \dfrac{d\xi}{da} = G = \alpha\dfrac{K^2}{E}$

The virtual work argument produces the extremely important relationship that the potential energy release rate per unit thickness, G, is given by:

$$G = \frac{K^2}{E} \text{ in plane stress} \tag{5}$$

and
$$G = \frac{K^2\left(1-v^2\right)}{E} \text{ in plane strain} \tag{6}$$

Critical values of G, i.e. the **fracture toughness**, are then identically equivalent to **critical values** of K, and it is these latter figures that are usually quoted for a material's fracture toughness.

Example 2e - Equivalence of G and K

Show that the stress analysis definition of $K = \sigma_{app}(\pi a)^{1/2}$ produces compatibility between Equations 2 and Equation 6 for plane strain.
What values of toughness, in terms of critical values of K, pertain to the low alloy steel and the maraging steel in Example 2a? How would this affect the comments made in that example?

Solution

Since
$$K = \sigma_{app}(\pi a)^{1/2}$$

and
$$G = \frac{K^2\left(1-v^2\right)}{E} \quad \text{(Equation 6)}$$

then
$$G = \frac{\sigma_{app}^{2}\pi a\left(1-v^2\right)}{E}$$

or
$$\sigma_{app} = \left(\frac{EG}{\pi\left(1-v^2\right)a}\right)^{1/2}$$

At fracture, G attains a critical value, G_{Ic}, and σ_{app} becomes σ_F. Hence,

$$\sigma_{app} = \left(\frac{EG_{Ic}}{\pi\left(1-v^2\right)a}\right)^{1/2} \quad \text{(as Equation 2)}$$

Alternatively, we may write;

$$K_c, K_{Ic} = \sigma_F\,(\pi a)^{1/2} \tag{7}$$

For Case (a) - Low Alloy Steel

(Work in MN and m)

In plane stress $\quad K_c^2 = EG_c \quad$ (Equation 5) $\; G_c = 24 \text{ kJ m}^{-2}$

$$= 200 \times 10^3 \times 24 \times 10^{-3}$$
$$= 4800$$

$$\therefore \; K_c = 69.3 \text{ MPam}^{1/2}$$

For Case (b) - Maraging Steel

$$G_c = 24 \text{ kJ m}^{-2}$$
$$K_c^2 = EG_c$$
$$\therefore \; K_c = 69.3 \text{ MPam}^{1/2}$$

> *Comment*
> Since the applied stress (from conventional design) is a constant fraction of the yield stress, the toughness measured as a critical K_{Ic} value, must increase in a linear fashion with yield stress to maintain constant defect tolerance.

2.4 Effects of Finite Boundaries

The expression for the stress distribution ahead of a sharp crack, Equation 3, holds only for an infinite body. If finite surfaces are present, the infinite plate solution is modified by an algebraic, trigonometric or polynomial function chosen to make the surface forces zero.

In other words, the resultant value of K, as the dominant term in a series expansion, is then not equal to $\sigma_{app}(\pi\,a)^{1/2}$, but is amended to take account of the width, W, of the testpiece or structure.

Examples of such K values are given in Table 2.1 for some of the more common loading configurations.

Type of Crack	Stress Intensity
Centre-crack, length 2a, in an infinite plate.	$K_I = \sigma_{app}(\pi a)^{1/2}$
Centre-crack, length 2a, in plate of width W.	$K_I = \sigma_{app}\left[W\tan\left(\dfrac{\pi a}{W}\right)\right]^{1/2}$ or $K_I = \sigma_{app}\left[\pi a \sec\left(\dfrac{\pi a}{W}\right)\right]^{1/2}$
Central, penny shaped crack, radius a, in infinite body.	$K_I = 2\sigma_{app}\left(\dfrac{a}{\pi}\right)^{1/2}$
Edge crack, length a, in semi-infinite plate.	$K_I = 1.12\sigma_{app}(\pi a)^{1/2}$
2 symmetrical edge-cracks each length a, in plate of total width W.	$K_I = \sigma_{app}W^{1/2}\left[\tan\left(\dfrac{\pi a}{W}\right)+0.1\sin\left(\dfrac{2\pi a}{W}\right)\right]^{1/2}$

Table 2.1 - Stress Intensity K Factors for Various Loading Geometries

Example 2f - Fracture in a Centre-Cracked Panel

A thick centre-cracked plate of a high strength aluminium alloy is 200 mm wide and contains a crack of length 80 mm. If it fails at an applied stress of 100 MPa, what is the fracture toughness of the alloy? What value of applied stress would produce fracture for the same length of crack in :-
(a) an infinite body
(b) a 120 mm wide plate?

Solution using tan function
From Table 2.1, for a centre-cracked plate,

$$K_I = \sigma_{app}\left[Wtan\left(\frac{\pi a}{W} \right) \right]^{1/2}$$

Since σ_{app} = 100 MPa, a = 0.04 m and W = 0.2 m

$$\frac{a}{W} = 0.2 \text{ and } tan\left(\frac{\pi a}{W} \right) = 0.726$$

Therefore, $\sigma_{app}[0.726]^{1/2}$ = 100 x 0.38
Hence, K_I = **38 MPa m$^{1/2}$**

This value will be taken as the plane strain fracture toughness, K_{Ic}, if the plate is very thick.

Case (a) - Infinite Body
From Table 2.1, $K_I = \sigma_{app}(\pi a)^{1/2}$

Thus, $38 = \sigma_{app}(0.04\pi)^{1/2}$

$$\therefore \sigma_F = \textbf{107.2 MPa}$$

Case (b) - Centre Cracked Plate
W = 0.12 m, $\frac{a}{W}$ = 0.333 and $tan\left(\frac{\pi a}{W} \right)$ = 1.732

From Table 2.1, $K_{Ic} = \sigma_F(1.732W)^{1/2}$ = 38

Thus, σ_F = **83.4 MPa**

Solution using sec function
From Table 2.1, for a centre-cracked plate,

$$K_I = \sigma_{app} \left[\pi a \sec\left(\frac{\pi a}{W} \right) \right]^{1/2}$$

since $\sigma_{app} = 100$ MPa, $a = 0.04$ m, $W = 0.2$ m

$$\frac{a}{W} = 0.2 \text{ and } \sec\left(\frac{\pi a}{W} \right) = 1.236$$

Therefore, $K_I = \sigma_{app}$ [0.04 x π x 1.236]$^{1/2}$ = 100 x 0.394

Hence, K_I = **39.4 MPa m$^{1/2}$** .

Case (a) - Infinite Body
From Table 2.1, $K_I = \sigma_{app}(\pi a)^{1/2}$

Thus, $39.4 = \sigma_F (0.04\pi)^{1/2}$

$$\therefore \ \sigma_F = \textbf{111.1 MPa}$$

Case (b) - Centre Cracked Plate
$W = 0.12$ m, $\dfrac{a}{W} = 0.333$ and $\sec\left(\dfrac{\pi a}{W} \right) = 2$

From Table 2.1, $K_{Ic} = \sigma_F(0.04\pi \text{ x } 2)^{1/2} = 39.4$

Thus, σ_F = **78.6 MPa**

Comment
It can be seen that the two formulations agree to within about 5%. The *sec* expression is regarded as the more accurate of the two.

Often, the calculations of the stress distribution ahead of a crack make use of stress functions which are written as a polynomial series, rather than single algebraic functions.

For example, in the single-edge notched (SEN) bend testpiece the stress intensity factor, K, may be derived from the expression:

$$K = \frac{PY_1}{BW^{1/2}} \tag{8}$$

where P is the applied load,

B is the testpiece thickness,

W is the testpiece width,

and Y_1 is a dimensionless polynomial function in odd half

powers of $\left(\dfrac{a}{W}\right)$ from $\left(\dfrac{a}{W}\right)^{1/2}$ to $\left(\dfrac{a}{W}\right)^{9/2}$

Practical details of this standard type of testpiece including tabulated values of Y_1, are given in Chapter 3 (Section 3.2).

Example 2g - Fracture in an SEN (Single Edge Notched) Bend Testpiece

A standard three-point-bend testpiece made of the aluminium alloy referred to in Example 2f, has a thickness, B, equal to 50 mm; a depth, W, of 100 mm; a loading span, $L = 2W = 200$mm, and is precracked by fatigue to give a total crack depth of 53 mm.
What applied load is required to produce fracture, assuming that linear elastic conditions apply?

Solution

$$P = \frac{KBW^{1/2}}{Y_1} \qquad \text{(from Equation 8)}$$

$W = 100$ mm, crack length, a $= 53$ mm $\therefore$ a/W $= 0.53$

for an a/W value of 0.53, Y_1 is found to be 11.74 (see Figure 3.1, p36)

From Example 2f, $K_{Ic} = 39.4$ MPa m$^{1/2}$

Then working in MN and m,

$$P = \frac{39.6 \times 0.05 \times (0.1)^{1/2}}{11.74}$$

$$= 0.0531 \text{ MN}$$

$$= 53.1 \text{ kN}$$

An alternative standard testpiece design is that of the **Compact Tension Specimen** (CTS), for which dimensions and Y-values are given in Figure 3.2. Although a given K-value can be obtained in a smaller amount of material, the loads required to give fracture (and hence the testing machine capacity) may have to be greater than for the SEN bend specimen.

Example 2h - Fracture in a CTS (Compact Tension) Testpiece

What load would be required to fracture a CTS specimen of dimensions $W = 100$ mm, $a = 53$ mm, $B = 50$ mm, made in the same aluminium alloy referred to in Examples 2f and 2g?

Solution

From the calibration of a CTS specimen (see Figure 3.2, p37)

$$P = \frac{KBW^{1/2}}{Y_2}$$

(9)

At fracture, $P_F = \frac{K_{Ic}BW^{1/2}}{Y_2}$

Now $a/W = 0.53$, so that $Y_2 = 10.62$ (see Figure 3.2)

Hence, $P_F = \frac{39.4 \times 0.05 \times (0.1)^{1/2}}{10.62}$

$$= 0.0589 \text{ MN}$$

$$= 58.9 \text{ kN}$$

2.5 Specimen Size Requirements

The assumed linear elastic stress analysis can be applied only if the extent of in-plane plasticity is small compared with testpiece dimensions. Additionally, the testpiece must be sufficiently thick in order that most of the deformation occurs under plane-strain conditions. If this is the case, the total fracture instability is coincident with the initiation of plane-strain fracture and the measured toughness is a true material property.
Experimentally, it has been found that this condition is met if:

$$B \geq 2.5 \left(\frac{K_{Ic}}{\sigma_y} \right)^2$$

where B is the thickness of the testpiece and σ_y is the 0.2% proof stress under the conditions of the test.
The in-plane dimensions, a and $(W-a)$ are set by two further considerations:
(i) To use the stress-intensity approach, the region of non-linear behaviour (i.e. plastic deformation) must be smaller than the region in which the K description of the stress field is a reasonable approximation. This implies that the plastic

zone size must be much smaller than the crack length, a. The rule set by the Standard (see Chapter Three) is that

$$a(\text{approx.}) \geq 2.5\left(\frac{K_{Ic}}{\sigma_y}\right)^2$$

(ii) If $(W-a)$ is decreased, holding B and a constant at a given stress intensity, the free surface of the back face of the specimen is brought closer to the plastic zone. This effectively reduces the shear stiffness of the elastic material in the uncracked ligament and the applied shear stress is able to produce an increased plastic zone size. This tends again to invalidate the K description of the stress field. The Standard specifies that

$$(W-a)\text{approx.} \geq 2.5\left(\frac{K_{Ic}}{\sigma_y}\right)^2$$

But, this could be an over-stringent requirement (ref. 19)

The net result of the above considerations is that the testpiece dimensions must be greater than those given by the Standard's specification, viz:

$$a, B, (W-a) \geq 2.5\left(\frac{K_{Ic}}{\sigma_y}\right)^2$$

$$\text{and } 0.45 < a/W < 0.55 \ (W = 2B) \tag{10}$$

Example 2i - Testpiece Dimensions (1)

It was assumed in Example 2g that a bend testpiece thickness of 50 mm was sufficient to guarantee plane strain fracture in a high-strength aluminium alloy,

(a) If the 0.2% proof stress of this alloy was found to be 450 MPa, in a uniaxial tensile test, would the above assumption be justified?

(b) What is the minimum strength level of an alloy of the same toughness that would still give a valid K_{Ic} result in this size of testpiece?

Solution

For (a) From Example 2f, $K_{Ic} = 39.4$ MPa m$^{1/2}$

From Equation 10, for a valid K_{Ic} test

$$B \geq 2.5\left(\frac{K_{Ic}}{\sigma_y}\right)^2$$

$$= 2.5 \left(\frac{39.4}{450} \right)^2$$

= 19.1 mm

A testpiece thickness of 50 mm is therefore more than adequate for a valid K_{Ic} result.

For (b) (Working in m)

$$B \geq 2.5 \left(\frac{K_{Ic}}{\sigma_y} \right)^2$$

$$0.05 \geq 2.5 \left(\frac{39.4}{\sigma_y} \right)^2$$

$$\sigma_y \geq 279 \text{ MPa}$$

The minimum value of proof stress is therefore **279 MPa.**

Example 2j - Testpiece Dimensions (2)

An alloy forging steel has a specified minimum proof stress, $\sigma_y = 800$ MPa and a guaranteed minimum fracture toughness, $K_{Ic} = 120$ MPa m$^{1/2}$.

(a) Calculate the minimum testpiece dimensions needed to carry out valid tests to check the toughness figures.

(b) Estimate the weights of sufficiently large standard SEN bend and CTS testpieces.

(c) Estimate the test machine capacity required.

Solution

For (a) $B \geq 2.5 \left(\frac{K_{Ic}}{\sigma_y} \right)^2$ from Equation 10

$$K_{Ic} = 120, \ \sigma_y = 800$$

$$\therefore B \geq 2.5 \ (0.15)^2 \text{ m}$$

$$= 56.25 \text{ mm}$$

and $W = 2B$

$$W = 112.5 \text{ mm}$$

For (b) *SEN bend specimen* (see Figure 3.1)

Testpiece length, $L = 4W + 10$ mm

$\therefore L = 460$ mm

Ignoring the notch, the volume is $B \times W \times L$

$\therefore$ volume $= 0.0029$ m^3
$\therefore$ mass $= 23$ kg (since density of steel ≈ 7900 kgm^{-3})

CTS specimen (see Figure 3.2)
length $= 2H = 1.2W = 135$ mm
total width $= 1.25W = 140.6$ mm
hole diameter, $D, = 0.25W = 28.125$ mm
Ignoring the notch, Volume $= (2H \times B \times C) - 2(\pi D^2/4 \times B)$
$$= (0.001068 - 0.000069) \text{ m}^3$$
$$= 0.001 \text{ m}^3$$
$\therefore$ mass $= 7.9$ kg

For (c) In order to obtain the minimum machine capacity required, assume the **minimum** permissible (a/W) value $= 0.45$ to calculate the **maximum** load required to give a K_{Ic} of 120 MPam$^{1/2}$ (cf Examples 2g and 2h).

SEN - bend specimen (see Figure 3.1)
From Equation 8,

$$P_F = \frac{K_{Ic}BW^{1/2}}{Y_1}$$

Now $W = 112.5$ mm, $B = 56.25$ mm and $Y_1 = 9.14$

$(a/W = 0.45)$

(Working in MN and m)

$$P_F = \frac{120 \times 0.05625 \times (0.1125)^{1/2}}{9.14}$$

$$= 0.248 \text{ MN}$$

The minimum machine capacity required is therefore **248 kN**

CTS Specimen (see Figure 3.2)

From equation 9,

$$P_F = \frac{K_{Ic}BW^{1/2}}{Y_2}$$

for $a/W = 0.45$, $Y_2 = 8.34$ (see Figure 3.2)

Hence   **P_F = 271 kN**

Comments
This example provides a good illustration of the practical features involved in the testing of a reasonably tough forging steel. The testpiece size becomes quite large and, for the SEN bend geometry, uncomfortably heavy for a single operator to handle easily. If the testing machine is of limited capacity, it may be necessary to resort to the bend geometry since the loading span can be increased to facilitate fracture. However, this may present difficulties in respect to the supply of material in suitable form.

If we take this example further, assuming the test machine is limited to 250 kN capacity, we could adopt a higher (a/W) value within the permitted range (i.e. up to $a/W = 0.55$) to lower the value of P_F. However, since a minimum guaranteed K_{Ic} value of 120 MPam$^{1/2}$ is specified, in practice a higher K_{Ic} will be the case, which in turn will require a higher value of P_F. Thus, B and W must be increased correspondingly.

In the present hypothetical case, B might be set at 65 mm ($W = 130$ mm) which would enable valid K_{Ic} results to be obtained up to a value of 129 MPam$^{1/2}$. If the test machine is limited to 250 kN, it would be just possible to test a standard testpiece, provided that $0.54 < a/W$ <0.55 ($Y > 12.15$), which would require exceptional control on the initial crack length. It would not be possible to fracture a CTS testpiece of 'valid' size in the machine. The use of higher a/W values, outside the Standard, is clearly attractive if the $(W-a)$ requirement is, indeed, over stringent.

2.6 Compliance Methods

The mathematical expressions for K become complicated for standard testpiece geometries and although the Y functions referred in Equation 8, are tabulated, the physical sense of what is being calculated is often lost.

For testpieces, a conceptually more direct method is to determine experimentally the change in stored elastic energy with increasing crack length, measured in terms of the displacement of the loading points, and to obtain a value of G from the expression:

$$G = \left(\frac{1}{B}\right)\frac{d\xi}{da}$$

Whilst examples of its usage are given below, the compliance approach has certain limitations. First, the measurement of small changes in displacement is difficult to make with accuracy. Additionally, in many structures, such as a pressure vessel, the position of the loading points is undefined. The theoretical methods must then be used.

Example 2k - Derivation of G Using Compliance Method

If the displacement-load relationship for a body of thickness, B, containing a crack of length a, is given by $u = CP$, where C is the compliance and is a function of a (a longer crack makes the specimen behave like a weaker elastic spring), show that the energy release rate per unit thickness, G, is given by.

$$G = \frac{P^2}{2B}\left(\frac{dC}{da}\right)$$

if the crack extends under a constant load P.

Solution
The strain energy stored in the body at crack length, a, is given by $\frac{1}{2} Pu = \frac{1}{2} CP$ (the area under the load-displacement curve).
If the crack extends at constant load, the displacement will increase to $(u+ du)$, whilst P remains constant, because the specimen with crack length $(a + da)$ behaves like a weaker spring.
The new strain energy stored is then increased to $\frac{1}{2} P(u + du)$.
But, the applied load does work of magnitude $-P(du)$, so that the net release of potential energy is equal to $-\frac{1}{2} Pdu$. i.e. $d\xi = -\frac{1}{2} Pdu$
Now $u = CP$

$$\therefore \frac{du}{da} = C\frac{dP}{da} + P\frac{dC}{da}$$

$$= P\frac{dC}{da} \text{ since } dP = 0 \text{ (constant load)}$$

Now, the change in energy for crack extension is given by:

$$\frac{d\xi}{da} = -\frac{1}{2}P\frac{du}{da}$$

$$\therefore \frac{d\xi}{da} = -\frac{1}{2} P^2 \frac{dC}{da}$$

G is defined as the positive value of $\frac{1}{B}\left(\frac{d\xi}{da}\right)$

Hence, $$G = \frac{P^2}{2B}\left(\frac{dC}{da}\right)$$ (11)

Example 21 - Compliance and Fracture Toughness

In the compliance calibration of an edge cracked fracture toughness testpiece of an aluminium alloy, it was observed that a load of 100 kN produced a displacement between the loading pins of 0.3000 mm when the crack length was 24.5 mm and 0.3025 mm when the crack length was 25.5 mm. The fracture load of an identical testpiece, containing a crack of length 25.0 mm is 158 kN. Calculate the critical value of the potential energy release rate at fracture and hence the plane-strain fracture toughness, K_{Ic}, of the alloy. All testpieces were 25 mm thick.

From Equation 6, for plane strain conditions, $G = \frac{K^2}{E}\left(1 - v^2\right)$ where

v has a value of 0.3 and E is 70 GPa for this alloy.

Solution

At $P = 100$ kN, $a_1 = 24.5$ mm and $u_1 = 0.3000$ mm

$a_2 = 25.5$ mm and $u_2 = 0.3025$ mm

Since $u = CP$

Then $C_1 = 0.3000 \times 10^{-2}$ mm kN^{-1}

$C_2 = 0.3025 \times 10^{-2}$ mm kN^{-1}

$$\frac{dC}{da} \approx \frac{C_2 - C_1}{a_2 - a_1}$$

$$\therefore \frac{dC}{da} = \frac{0.0025 \times 10^{-2}}{1} \, kN^{-1}$$

$$= 2.5 \times 10^{-5} \, kN^{-1}$$

From Equation 11,

$$G_{crit} = \frac{P_F^2}{2B}\left(\frac{dC}{da}\right)$$

$$= \frac{158^2 \times 2.5 \times 10^{-5}}{2 \times 25}$$

$$= 12.5 \ kJm^{-2}$$

From Equation 6,

$$K_{Ic}{}^2 = \frac{G_{Ic} E}{1 - v^2}$$

$$= \frac{12.5 \times 10^{-3} \times 70 \times 10^3}{0.91}$$

$$= 962$$

$$\therefore K_{Ic} = 31.0 \ MPa \ m^{1/2}$$

2.7 Weight Functions

We have shown (example IV) that the potential energy release rate per unit thickness, G, is given by

$$G = \alpha K^2 / E \qquad \text{(equations 4 and 5)}$$

and that the work done on extending a crack is given (example 2d, p9) by

$$\delta \xi = \int_0^{\delta a} \sigma u \, dr. \tag{12}$$

For a crack growing from zero to a crack length of a, r becomes identical to x ; hence $dr = dx$ and we have

$$d\xi = \int_0^a \sigma u \, dx. \tag{13}$$

Now

$$\frac{d\xi}{da} = \int_0^a \sigma \frac{du}{da} dx \tag{14}$$

since neither σ nor dx vary with a, and as G is defined as $d\xi/da$ per unit thickness, then

$$\frac{\alpha K^2}{E} = \int_0^a \sigma \frac{du}{da} dx \tag{15}$$

This may be rearranged this to give the form:

$$K = \int_0^a \frac{\sigma E}{\alpha K} \frac{du}{da} dx \tag{16}$$

Define $m(x)$ as

$$m(x) = \frac{E}{\alpha K} \frac{du}{da} \tag{17}$$

hence

$$K = \int_0^a \sigma\, m(x)\, dx.$$
(18)

The term $m(x)$ is known as the **weight function**.

Example 2m - Derivation of a Weight Function for a Crack in an Infinite Sheet

Using the equation for u_2 (equation A22 in the Appendix) derive the weight function for a crack of length $2a$ in an infinite sheet under an applied stress of σ_{app}.

<u>Solution</u>
From equation A22 we have

$$u_2 = \frac{2}{E}\left(1-v^2\right)\sigma_{app}\sqrt{a^2-x^2}.$$

As $\alpha = (1 - v^2)$ this becomes

$$u_2 = \frac{2}{E}\alpha\sigma_{app}\sqrt{a^2-x^2}$$

and

$$\frac{du}{da} = \frac{2\alpha}{E}\sigma_{app}\frac{a}{\sqrt{a^2-x^2}}.$$

Note that we are differentiating with respect to a not with respect to x. This can be substituted into the equation for the weight function $m(x)$ to obtain

$$m(x) = \frac{E}{\alpha K}\frac{2\alpha}{E}\sigma_{app}\frac{a}{\sqrt{a^2-x^2}}$$

$$= \frac{2}{K}\sigma_{app}\frac{a}{\sqrt{a^2-x^2}}.$$

From table 2.1 we have $K = \sigma_{app}\sqrt{(\pi a)}$ for an infinite sheet. Substituting for K we obtain

$$m(x) = \sqrt{\frac{a}{\pi}}\frac{2}{\sqrt{a^2-x^2}}$$

Although for this case it is easier to use the formula found in table 2.1, the weight function method becomes useful for more complex loadings and geometries. The function $m(x)$ is a geometric function which depends simply on the crack length and shape and on the geometry of the testpiece or component in which the crack is present. A number of weight functions for different geometries are available in the literature (e.g. ref 20), and it can be of value to apply a weight function known for one geometry to a component with a roughly comparable geometry but for which the weight function is unknown.

Example 2n - Calculation of the weight function for a cracked bend testpiece

A bend testpiece in pure bending (four point loading) contains a crack of length a. The stress across the testpiece varies linearly from a maximum tensile stress (σ_m) at one surface, through zero in the centre, to a maximum compressive stress ($-\sigma_m$) at the other surface. The stress intensity factor is given by

$$K = 1.1215 \frac{2}{\pi} (\pi a)^{1/2} \int_0^a \frac{\sigma(x)\,dx}{(a^2 - x^2)^{1/2}}$$

where x is the distance from the top surface of the testpiece. Calculate the value of f in the equation $\qquad K = f \sigma_m (a\pi)^{1/2}$ for the bend testpiece if the crack is 0.05 a/W in length.

Solution
The equation for the variation of the stress within the testpiece must be derived. As stress drops linearly from σ_m at $x = 0$, to zero at $x = W/2$ and on to $-\sigma_m$ at $x = W$, the variation of the stress with distance, x, in the uncracked body, is

$$\sigma(x) = \sigma_m (1 - 2x/W)$$

The equation above becomes

$$K = 1.1215 \frac{2}{\pi} (\pi a)^{1/2} \sigma_m \int_0^a \frac{(1 - 2x/W)}{(a^2 - x^2)^{1/2}}\,dx$$

$$K = 1.1215 \frac{2}{\pi} (\pi a)^{1/2} \frac{\sigma_m}{W} \left\{ \int_0^a \frac{W}{(a^2 - x^2)^{1/2}}\,dx - \int_0^a \frac{2x}{(a^2 - x^2)^{1/2}}\,dx \right\}$$

For $\qquad \int_0^a \frac{1}{(a^2 - x^2)^{1/2}}\,dx$

let $x = a \sin\omega$, then $dx = a \cos\omega\,d\omega$. When $x = 0$, $\omega = 0$ and when $x = a$, $\omega = \pi/2$.

Hence $\qquad \int_0^a \frac{1}{(a^2 - x^2)^{1/2}}\,dx = \int_0^{\pi/2} \frac{a \sin\omega}{(a^2 - a^2 \cos^2\omega)^{1/2}}\,d\omega$

$$= \int_0^{\pi/2} \frac{\sin\omega}{(1 - \cos^2\omega)^{1/2}}\,d\omega$$

As $\sin^2\omega = 1 - \cos^2\omega$, then

$$\int_0^a \frac{1}{(a^2 - x^2)^{1/2}} dx = \int_0^{\pi/2} \frac{\sin \omega}{(\sin^2 \omega)^{1/2}} d\omega = \int_0^{\pi/2} d\omega = [\omega]_0^{\pi/2} = \frac{\pi}{2}$$

For $\quad \int_0^a \frac{2x}{(a^2 - x^2)^{1/2}} dx$

let $u^2 = a^2 - x^2$, then $dx = -u/x\ du$. When $x = 0$, $u = a$ and when $x = a$, $u = 0$.

Hence

$$\int_0^a \frac{2x}{(a^2 - x^2)^{1/2}} dx = -\int_a^0 \frac{2u}{(u^2)^{1/2}} du = -2\int_a^0 du = -2[u]_a^0 = 2a$$

Substituting back into the equation for K, we obtain

$$K = 1.1215 \frac{2}{\pi} (\pi a)^{1/2} \frac{\sigma_m}{W} \left\{ \frac{W\pi}{2} - 2a \right\}$$

For a crack length of 0.05 a/W

$$K = 1.1215 \frac{2}{\pi} (\pi a)^{1/2} \sigma_m \left\{ \frac{\pi}{2} - 2 \times 0.05 \right\}$$

$$K = 1.050\ \sigma_m\ (\pi\ a)^{1/2}$$

Thus, $f = 1.050$

Comments

The weight function derived here is 1.050 whereas that given in the standard reference 20 is 1.071. This is because the equation used assumes the form for an edge crack in a semi-infinite sheet. However, the error in this method is only 5% up to a crack length of 0.075 *a/W*. It may be noted that the stress intensity factor calculated from equation 8 and a compliance function such as that in reference 21 differs by only 1% from that calculated by the weight function method.

The weight function method is very useful for the cases where there are stress raisers or residual stresses in a material as they allow the modified stress intensity factor to be calculated.

Example 2o - Comparison of the Stress Intensity Factors of a Crack and a Crack Emanating from a Hole in an Infinite Sheet.

The stress intensity factor ahead of one of two cracks emanating from opposite sides of a hole in an infinite sheet under uniaxial tension is given by the equation

$$K = f\,\sigma_{app}\sqrt{\pi a}$$

where a is the crack length from the edge of the hole.
The value of f is given in the graph below. Note that, as $a/_R \to 0$, so
$f \to 1.12 \times 3$, the factor for an edge crack, Table 2.1, multiplied by
the elastic stress concentration factor for a hole, 3.

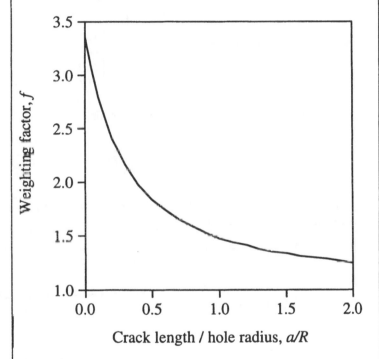

Weighting factor, f (vertical axis)

Crack length / hole radius, a/R (horizontal axis)

Figure 2.4

If the hole has a radius, R, show that the hole and cracks can be
represented by a single crack with a maximum error of 5% for
cracks longer than 0.1 a/R.

<u>Solution</u>
The stress intensity factor for a single crack of length $2a$, is given
in Table 2.1 as

$$K = \sigma_{app}\sqrt{\pi a}$$

To compare the two formulae the crack length of the single crack
becomes $2a + 2R$ and hence the equation becomes

$$K = \sigma_{app}\sqrt{\pi(a+R)}$$

Figure 2.4 gives the crack length as a multiple of the hole radius, so that the equation is best modified to the form

$$K = \sigma_{app}\sqrt{r}\sqrt{\pi\left(\frac{a}{R}+1\right)}$$

and for the hole and cracks,

$$K = f\,\sigma_{app}\sqrt{r}\sqrt{\pi\left(\frac{a}{R}\right)}.$$

If we divide one equation by the other we obtain

$$\frac{K_{hole}}{K_{crack}} = f\,\frac{\sqrt{\dfrac{a}{R}}}{\sqrt{\dfrac{a}{R}+1}}$$

Which when plotted against the ratio of crack length to hole radius, *a/R*, gives the graph

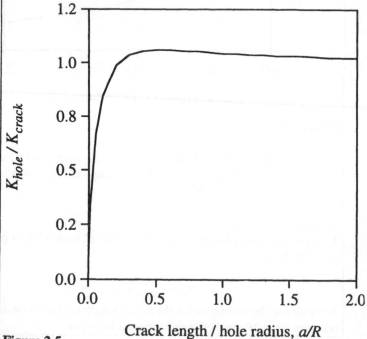

Figure 2.5

The sharp crack can be seen to over-estimate the stress intensity factor due to a hole plus cracks up to a crack length of 0.1 of the hole radius. The hole and cracks then give a stress intensity factor up to 5% higher (at *a/R* of 0.5) before asymptoting to the value for a single crack.

Weight function methods can also be used to calculate the effect of residual stresses around a particle

Example 2p - Residual Stresses Around a Particle

A particle of radius 2 μm in a large sheet of steel induces a tensile residual stress field on the surrounding matrix due to thermal mismatch. At the particle matrix interface this stress is 1000 MPa. Calculate the effect of this residual stress on the stress intensity factor of two cracks each of length 4 μm emanating from the centre of this particle, if the sheet is under a uniaxial tension of 200 MPa. The stress intensity factor is given by the equation

$$K = f \sigma_{res} \sqrt{\pi a}$$

The length of each crack, 4 μm, is twice the particle radius, and *f* is given as 0.2511 for a crack of half length 2*R* (where *R* is the particle radius).

Solution
The stress intensity factor due to the residual stress is
$$K = f \sigma_{res} \sqrt{\pi a}$$
$$K = 0.2511 \times 1000 \sqrt{\pi 4 \times 10^{-6}}$$
$$K = f \sigma_{res} \sqrt{\pi a}$$
$$K = 0.89 \text{ MPa m}^{1/2}.$$
The stress intensity factor due to a hole and a crack given in Example 2o is
$$K = f \sigma_{app} \sqrt{\pi a}$$
and *f* is taken from figure 2.4, for a crack / hole radius of 2, as 1.25. This gives
$$K = 1.25 \times 200 \sqrt{4\pi \times 10^{-6}}$$
$$K = 0.89 \text{ MPa m}^{1/2}.$$
The stress intensity factor due to the residual stress is equal to that of the applied stress and thus doubles the stress intensity factor.

> **Comment.**
> If the applied stress approaches 1000 MPa then the stress intensity
> factor due to the residual stress falls to 20% of that of the applied
> stress. Although these values of stress intensity factor are much
> less than the fracture toughness of steel, anything which acts as a
> stress raiser in this way will have a marked effect on the fatigue
> life (see Chapter 5) because the fatigue life depends on the stress
> intensity factor range and, in this example, this is increased by
> increasing the maximum stress.

2.8 Effects of Microstructure

The microstructure within a material can have marked effects upon the
mechanical properties of the material. Here are two examples; the Petch
equation and transformation toughening.

The Petch equation gives an expression for the lower yield stress, relating to
conditions at the front of a Luders band, propagating along a tensile testpiece.
The physical situation which is modelled is that of a slip-band in one grain,
impinging on the grain-boundary between this and an unyielded grain (figure
2.6).

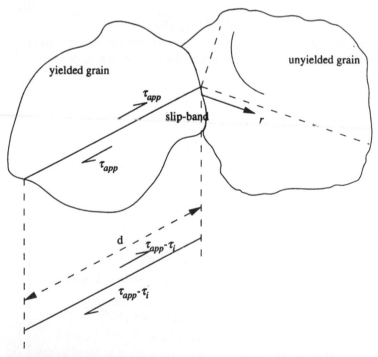

Figure 2.6

The dislocations in the slip-band have to move against a "frictional stress", τ_i. The slip-band is modelled as a Mode II shear crack, acted on by a reduced shear stress.

$$\tau_{eff} - \tau_{app} - \tau_i$$

The local stress at a distance, r, ahead of the tip of the crack, is given by the form

$$\tau_{(r)} = \left(\tau_{app} - \tau_i\right)(d/4r)^{1/2}$$

This derives from the expression for the stress intensity factor pertaining to a central crack of length $2a = d$, subjected to a Mode II (shear) stress, τ. The general form is:

$$K = \tau(\pi a)^{1/2}$$
$$|\tau_{(r)} = \tau(\pi a/2\pi r)^{1/2}$$
$$= \tau(d/4r)^{1/2}$$

This applied stress produces a value

$$K_{app} = \tau_{app}(d/4r)^{1/2}$$

and this is reduced by the friction stress by an amount

$$\tau_{(r)} = \left(\tau_{app} - \tau_i\right)(d/4r)^{1/2}$$

Hence
$$\tau_{(r)} = \left(\tau_{app} - \tau_i\right)(d/4r)^{1/2}$$

It is now supposed that yield is produced in the unyielded grain when $\tau_{(r)}$ attains a critical value τ^* at a critical distance r^* ahead of the tip of the slip-band. For lightly-pinned dislocations, τ^* would be the unpinning stress and r^* would be the average distance to the nearest dislocation source.

We then have
$$\tau^* = \left(\tau_{app} - \tau_i\right)(d/4r^*)^{1/2}$$

or by rearrangement
$$\tau_{app(Y)} = \tau_i + \left[(4r)^{1/2}\,\tau^*\right]d^{-1/2}$$

The term $\left[(4r^*)^{1/2}\tau^*\right]$ is written as k_Y^s where the superscript 's' denotes shear.

To convert from shear to tension, both sides of the equation are multiplied by the `Taylor factor', M, which takes account of the statistical distribution of available slip systems in different, randomly-oriented crystal systems. For b.c.c. crystals, $M = 2.75$, for the f.c.c. system, $M = 3.2$.

Hence,
$$\sigma_y = M\tau_{app(Y)} = \sigma_i + k_Y\, d^{-1/2}$$

This is the Petch equation, in which all terms refer to uniaxial tension. The temperature dependence of the yield stress depends on those of σ_i and k_Y. The friction stress, σ_i, generally contains two components: one, `athermal', deriving from strong barriers, such as dislocation networks, precipitates or irradiation damage; the other, temperature-dependent, deriving from weak barriers, such as isolated solute atoms or the Peierls-Nabarro stress. The temperature dependence of k_Y relates to the strength of pinning of dislocation sources by interstitial solute atoms: weak pinning gives temperature dependence, but strong pinning gives rise to a temperature-independent k_Y (for very strong pinning, new slip dislocations may be created athermally at the common grain boundary).

Example 2q - The Effect of Grain Size on Yield Stress

Given that the unpinning stress for weakly-pinned dislocations is 125 MPa and that the nearest dislocation source is located at a distance of 1 μm ahead of the slip-band tip, calculate the values of tensile yield stress for randomly oriented iron polycrystals of mean grain size 100 μm and 10 μm, having a friction stress of 25 MPa. What value of k_Y is obtained?

Solution

Values of τ^* and r^* are given as 125 MPa and 1 μm respectively.

Hence, for $d = 100$ μm, (in MN and m)

$$\tau_{app(Y)} = \tau_i + \left[(4r^*)^{1/2} \tau^* \right] d^{-1/2}$$

$$\tau_{app(Y)} = 25 + \left[\left(4 \times 10^{-6} \right)^{1/2} 125 \right] \left(10^{-4} \right)^{-1/2}$$

$$\tau_{app(Y)} = 25 + 25$$

$$\tau_{app(Y)} = 50 \text{ MPa}$$

Now iron is b.c.c.; hence $M = 2.75$

$$\sigma_y = M \tau_{app(Y)} = 2.75 \times 50$$

$$\sigma_y = 137.5 \text{ MPa}$$

Now $$k_Y = (4r^*)^{1/2} \tau^*$$

hence $$k_Y = 0.25 \text{ MPa m}^{1/2}$$

For $d = 10$ μm $$\tau_{app(Y)} = \tau_i + \left[(4r^*)^{1/2} \tau^* \right] d^{-1/2}$$

$$\tau_{app(Y)} = 25 + \left[\left(4 \times 10^{-6} \right)^{1/2} 125 \right] \left(10^{-5} \right)^{-1/2}$$

$$\tau_{app(Y)} = 25 + 79.06$$

$$\tau_{app(Y)} = 104.06 \text{ MPa}$$

Now iron is b.c.c.; hence $M = 2.75$

$$\sigma_y = M\,\tau_{app}(Y) = 2.75 \times 104.06$$
$$\sigma_y = 286.16\ \text{MPa}$$

Now $\quad k_Y = (4r*)^{1/2}\,\tau*$

hence $\quad k_Y = 0.25\ \text{MPa m}^{1/2}$

The yield stress is increased on decreasing the grain size but the value of k_Y is independent of grain size.

Transformation toughening effects the fracture toughness of a material by causing a clamping stress on the crack which opposes crack opening. When stressed, a material (e.g. zirconia-containing ceramics) may undergo a phase change. This phase change will usually be in the highly stressed area, ahead of a crack tip, and will usually involve a volume change. This volume change can exert a pressure on the crack which opposes opening and hence increases the fracture toughness.

Example 2r - The Effect of Transformation Toughening on Fracture Toughness.

A model of transformation toughening in ceramics assumes that the effect of the transformation is to exert a clamping pressure, p, over a short distance immediately behind the crack tip. Calculate the improvement in toughness predicted by this model for a crack of half-length 1 mm in a ceramic of surface energy 1 Jm^{-2} if the transformed region is 100 μm long and produces a clamping pressure of 60 MPa. Young's modulus for the ceramic is 400 GPa and Poisson's ratio is 0.25.

Solution
The stress intensity for a crack of half-length a loaded by a pressure p over a distance $(a-c)$ from the tip of the crack can be calculated from equation 18:

$$K = \int_{a-c}^{a} \sigma\,m(x)\,dx$$

and m(x) is given by (Example2m)

$$m(x) = \sqrt{\frac{a}{\pi}}\,\frac{2}{\sqrt{a^2 - x^2}}$$

This gives

$$K = 2\sqrt{\frac{a}{\pi}}\,p\int_{a-c}^{a}\frac{dx}{\sqrt{a^2 - x^2}}$$

If $x = a \sin \theta$, then $dx = a \cos \theta \, d\theta$.

$$\int_{a-c}^{a} \frac{dx}{\sqrt{a^2 - x^2}} = \int_{\sin^{-1} c/a}^{\sin^{-1} 1} \frac{a \cos \theta \, d\theta}{\sqrt{a^2 - a^2 \sin^2 \theta}}$$

$$= \int_{\sin^{-1} c/a}^{\sin^{-1} 1} \frac{\cos \theta \, d\theta}{\cos \theta} = \int_{\sin^{-1} c/a}^{\sin^{-1} 1} d\theta$$

Therefore the stress intensity due to the transformation is (in MN and m)

$$K = 2p \left(\frac{a}{\pi} \right)^{1/2} \left\{ \sin^{-1} 1 - \sin^{-1} \left(\frac{c}{a} \right) \right\}$$

$$K = 2 \times 60 \left(\frac{0.001}{\pi} \right)^{1/2} \left\{ \sin^{-1} 1 - \sin^{-1} \left(\frac{0.0009}{0.001} \right) \right\}$$

$$K = 0.966 \text{ MPam}^{1/2}$$

From equations 1 and 2 we have

$$G_{Ic} = 2\gamma$$
$$G_{Ic} = 2 \text{ Jm}^{-2}$$

From equations 5 and 6 we have

$$G_{Ic} = \frac{\alpha K_{Ic}^2}{E}$$

For an untransformed material K_{Ic} is given by

$$K_{Ic} = \sqrt{\frac{G_{Ic} E}{\alpha}}$$

$$= \sqrt{\frac{2 \times 400 \times 10^9}{1 - 0.25^2}}$$

$$= 0.924 \text{ MPam}^{1/2}$$

The addition of the stress intensity factor for the toughening leads to

$$K_{Ic} = 0.966 + 0.924$$
$$K_{Ic} = 1.89 \text{ MPam}^{1/2}$$

Now

$$G_{Ic} = \frac{\alpha K_{Ic}^2}{E}$$

$$G_{Ic} = \frac{(1 - 0.25^2)(1.89 \times 10^6)^2}{400 \times 10^9}$$

$$G_{Ic} = 8.37 \text{ Jm}^{-2}$$

Comment
It can be seen that the stress intensity at failure is doubled by the transformation but the fracture toughness is increased fourfold.

3. Fracture Toughness Determination

The basic aim of the fracture toughness test is to obtain a reproducible value of the critical toughness of the material, denoted by K_{Ic}, for mode I opening.

The object in this Section is to go briefly through the steps required for the experimental determination of K_{Ic}. For more precise information and guidance, reference should be made to BS 7448 'Methods for Determination of K_{Ic}, Critical CTOD and Critical J Values of Metallic Materials'.

3.1 Outline of Test

The test involves loading a test-piece which contains a pre-existing crack usually developed by fatigue from a machined notch. The practical objective is to determine the applied force at which a given amount of crack extension has taken place from the pre-existing crack. This information is established from a force / displacement test record in terms of a given deviation from linearity. The plane-strain fracture toughness, K_{Ic}, is then calculated using the stress analysis relationship developed specifically for the particular type of test specimen involved.

3.2 General Testing Details

In order to determine the fracture toughness K_{Ic} value from a laboratory test it is necessary to know the applied force, the initial crack length and the way in which the K value varies with increasing crack length, as a result of the boundary and loading conditions imposed by the particular design of test specimen - otherwise referred to as the *K-COMPLIANCE* or *K-CALIBRATION* function.

As was mentioned in Chapter 2, K-calibrations can be derived experimentally for a given specimen design (see Example 2l) or mathematically, the latter procedure being the more accurate.

Two standard types of testpiece are currently recommended in BS 7448 (and in ASTM E399-74T). The first is a single-edge-notched (SEN) bend testpiece and the second, a compact tension (CTS) testpiece. These are illustrated in Figure 3.1 and Figure 3.2 respectively, together with tables of K-calibration values (i.e. Y functions in Equation 8 and Equation 9).

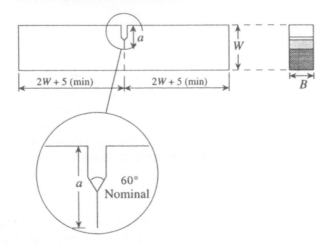

Width   = W
Thickness   = $B = 0.5W$
Half Loading Span L = $2W$
Effective crack length a = $0.45W$ to $0.55W$
Dimensions in mm.

Values of Y_I as a function of a/W for bend testpiece, three point loaded, with half-loading span to testpiece width ratio 2:1.

a/W	0.000	0.001	0.002	0.003	0.004	0.005	0.006	0.007	0.008	0.009
	Stress Intensity factor coefficient, Y_I									
0.450	9.14	9.17	9.20	9.22	9.25	9.28	9.31	9.33	9.36	9 39
0.460	9.42	9.45	9.47	9.50	9.53	9.56	9.59	9.62	9.65	9.68
0.470	9.70	9.73	9.76	9.79	9.82	9.85	9.88	9.91	9.94	9.97
0.480	10.01	10.04	10.07	10.10	10.13	10.16	10.19	10.22	10.26	10.28
0.490	10.32	10.35	10.38	10.42	10.45	10.48	10.52	10.55	10.58	10.62
0.500	10.65	10.68	10.72	10.75	10.78	10.82	10.86	10.89	10.93	10.96
0.510	11.00	11.03	11.07	11.10	11.14	11.18	11.21	11.25	11.29	11.32
0.520	11.36	11.40	11.43	11.47	11.51	11.55	11.59	11.63	11.66	11.70
0.530	11.74	11.78	11.82	11.86	11.90	11.94	11.98	12.02	12.06	12.10
0.540	12.15	12.19	12.23	12.27	12.31	12.35	12.40	12.44	12.48	12.53
0.550	12.57	------	------	------	------	------	------	------	------	------

Figure 3.1

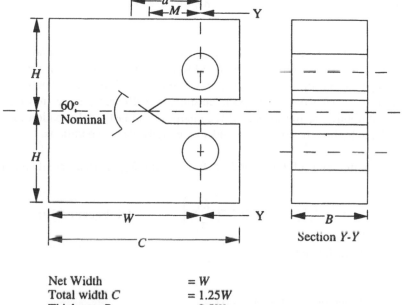

Net Width $= W$
Total width C $= 1.25W$
Thickness B $= 0.5W$
Half height H $= 0.6W$
Hole diameter $= 0.25W$
Effective notch length M $= 0.25W$ to $0.40W$
Effective crack length a $= 0.45W$ to $0.55W$
Dimensions in mm.

Values of Y_2 as a function of a/W for a compact tension testpiece

a/W	0.000	0.001	0.002	0.003	0.004	0.005	0.006	0.007	0.008	0.009
	Stress Intensity factor coefficient , Y_2									
0.450	8.34	8.36	8.39	8.41	8.43	8.46	8.48	8.51	8.53	8.56
0.460	8.58	8.60	8.63	8.65	8.68	8.70	8.73	8.75	8.78	8.80
0.470	8.83	8.86	8.88	8.91	8.93	8.96	8.99	9.01	9.04	9.07
0.480	9.09	9.12	9.15	9.17	9.20	9.23	9.26	9.29	9.31	9.34
0.490	9.37	9.40	9.43	9.45	9.48	9.51	9.54	9.57	9.60	9.63
0.500	9.66	9.69	9.72	9.75	9.78	9.81	9.84	9.87	9.90	9.93
0.510	9.96	10.00	10.03	10.06	10.09	10.12	10.16	10.19	10.22	10.25
0.520	10.29	10.32	10.35	10.39	10.42	10.45	10.49	10.52	10.56	10.59
0.530	10.63	10.66	10.70	10.73	10.77	10.80	10.84	10.87	10.91	10.95
0.540	10.98	11.02	11.06	11.10	11.13	11.17	11.21	11.24	11.29	11.33
0.550	11.36	------	------	------	------	------	------	------	------	------

Figure 3.2

In simple terms, the fracture toughness (K_{Ic}) is given by

$$K_{Ic} = \frac{PY}{BW^{1/2}}$$

(19)

As was explained in Section 2.5, before a valid K_{Ic} result can be obtained in the laboratory, it is necessary to prescribe limits in terms of specimen size and crack length so that there is sufficient constraint to achieve fully plane-strain conditions. In other words, the specimen must be sufficiently large in section for the contribution to the measured toughness from plastic deformation to be negligible, so that fracture proceeds under essentially elastic conditions.

Current requirements of the Standard for a valid K_{Ic} measurement are as follows:

$$B \geq 2.5 \left(\frac{K_{Ic}}{\sigma_y} \right)^2$$

(cf. Equation 10)

Where a = crack length

B = specimen thickness

W = specimen width $(W = 2B)$

and $0.45 < a/W < 0.55$

σ_y = 0.2% proof stress under the conditions of test.

It is impossible to determine at the outset of testing whether a valid K_{Ic} measurement has been made. It is first necessary to calculate a provisional result, K_Q, which involves a geometrical construction on the force / displacement test record and then to determine whether this result is consistent with the testpiece requirements, as specified above (see Standard for precise details).

3.3 Derivation of K_Q and K_{Ic} - The 'Offset Procedure'

Referring to Figure 3.3, which illustrates the main types of force / displacement test records, draw the secant line OP_5 through the origin with a slope of 5% less than the slope of the tangent OA to the initial part of the record. P_5 is the lowest force at the intersection of, or tangency to, the secant with the force / displacement record. The force P_Q is equal to P_5, or any higher force which precedes P_5, and this value is used to calculate K using Equation 19, with the value of Y taken from the appropriate K-calibration tables (see Figures 3.1 and 3.2)

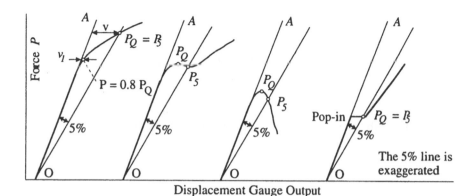

Displacement Gauge Output

Figure 3.3

The value of P_{max}/P_Q is recorded. If this ratio exceeds 1.10 it is likely that K_Q bears insufficient relation to K_{Ic} and the curve should be rejected.

The factor $2.5(K_Q/\sigma_y)^2$ is next calculated and if this is less than both the thickness and crack length of the testpiece and all other validity criteria are met, then K_Q is equal to K_{Ic} (see Equation 10). Otherwise it is necessary to use a larger testpiece to determine K_{Ic} , such that the thickness and crack length are not less than $2.5(K_{Ic}/\sigma_y)^2$.

Example 3a - K_{Ic} Determination (Bend Testpiece)

Figure 3.4 shows the force / displacement record obtained on testing a 25mm thick high strength low alloy steel in 3-point bending with a 4:1 loading span. The sketch shows the fractured surface of the test specimen. Determine whether the test represents a valid K_{Ic} result.
It may be assumed that the fatigue cracking, testing and measurement are in accordance with BS 7448.
The 0.2% proof stress is 1640 MPa.

<u>Solution</u>
From the testpiece measurements,
$a = 23.75$ mm, $B = 25.00$ mm, $W = 50.00$ mm
$\quad\quad$ \ $a/W = 0.475$ (ie $0.45 < a/W < 0.55$)
The compliance function (Y_1) is obtained from Figure 3.1
i.e. $\quad\quad\quad\quad\quad\quad\quad$ $Y_1 = 9.85$ (note *half* loading span is 2:1)

From the test record, the 5% secant construction provides

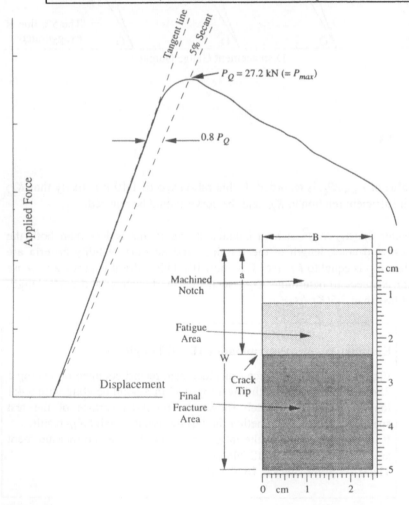

Figure 3.4

$$P_Q = 27.2 \text{ kN}$$

(The P_{max}/P_Q requirement is met)

Hence, from Equation 8, (working in MN and m)

$$K_Q = \frac{P_Q Y_1}{BW^{1/2}}$$

i.e.
$$K_Q = \frac{0.0272 \times 9.85}{0.025 \times (0.050)^{1/2}}$$

$$= 47.9 \text{ MPa m}^{1/2}$$

From Equation 10,

$$2.5\left(\frac{K_Q}{\sigma_y}\right)^2 = 2.5\left(\frac{47.9}{1640}\right)^2$$

$$= 2.1 \text{ mm}$$

For a valid K_{Ic} result

$$a, B \geq 2.5\left(\frac{K_Q}{\sigma_y}\right)^2$$

i.e. $\qquad a, B \geq 2.1 \text{ mm}$

From the testpiece measurements, this is seen to be the case.

Hence, $\qquad K_Q = K_{Ic} = 47.9 \text{ MPa m}^{1/2}$

Example 3b - K_{Ic} Determination (Tension Testpiece)

Figure 3.5 shows the force / displacement record obtained on testing a 50mm CTS tension testpiece taken from a forging of proof stress 1050 MPa. A sketch of the fractured testpiece is shown.
Determine whether the test represents a valid K_{Ic} result.

Solution

From the testpiece measurements,

$a = 52.09$ mm, $B = 49.93$ mm, $W = 100.03$ mm

$\therefore a/W = 0.521$ (i.e. $0.45 < a/W < 0.55$)

The compliance function (Y_2) is obtained from Figure 3.2.

i.e. $Y_2 = 10.32$

From the test record, the 5% secant construction provides

$$P_Q = 0.241 \text{ MN}$$

Also, $P_{max} = 0.261 \text{ MN}$

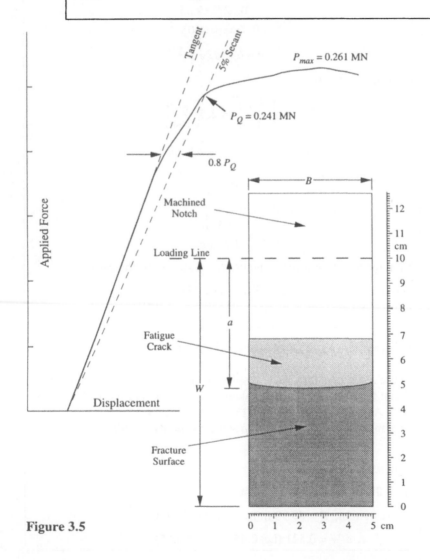

Figure 3.5

so that
$$\frac{P_{max}}{P_Q} = \frac{0.261}{0.241} = 1.08$$

From Equation 9 (working in MN and m)

$$K_Q = \frac{P_Q Y_2}{BW^{1/2}}$$

$$= \frac{0.241 x 10.32}{0.04993 x (0.100)^{1/2}}$$

$$= 158 \text{ MPa m}^{1/2}$$

From Equation 10,

$$2.5\left(\frac{K_Q}{\sigma_y}\right)^2 = 2.5\left(\frac{158}{1050}\right)^2$$

For a valid K_{Ic} result

$$a, B \geq 2.5\left(\frac{K_Q}{\sigma_y}\right)^2$$

i.e. $\geq 56 \text{ mm}$

From the testpiece measurements of a and B, this is seen not to be the case.

Hence, the test result is *invalid* and must be considered only as a K_Q value.

Comment
Although this test result is invalid according to the $2.5(K/\sigma_y)^2$ criteria, the P_{max}/P_Q ratio of 1.08 means that the test result should not be too far removed from K_{Ic}. The K_{Ic} measurement capacity of this 50mm thick testpiece is given by:

$$K_{Ic} = \left(\frac{B}{2.5}\right)^{1/2} \sigma_y \qquad \text{(from Equation 10)}$$

i.e. $\left(\frac{0.050}{2.5}\right)^{1/2} 1050 = 148 \text{ MPa m}^{1/2}$

To achieve a valid K_{Ic} result from this material it would be necessary to increase the specimen dimensions only slightly.

3.4 Application of K_{Ic}

It may not be appreciated that, in practice, there is no such thing as a 'defect-free' material. Cracks or crack-like defects are inherent in every component. The important point to ensure is that these cracks remain harmless and do not grow in service (i.e. by overstressing, fatigue, stress corrosion etc.) to a critical size.

The adoption of fracture mechanics is particularly useful in this respect, since it enables us to evaluate the reliability of a structure or component in service. Its use in design can help in the correct choice of material and to judge whether a defect is likely to be of a size to cause catastrophic fracture.

In simple terms, the higher the value of K_{Ic}, the greater is the resistance of the material to catastrophic failure.

In applying fracture toughness methods to design, we determine the critical stress intensity factor, K_{Ic}, at fracture in a relatively simple laboratory testpiece, where the relationship between applied stress and crack size is known from a suitable compliance analysis (see Section 3.1 to 3.3).

From a knowledge of K_{Ic}, it should be possible to calculate the applied stress to cause failure, for a given defect size, or, alternatively, to predict the size of defect necessary to cause failure for a given applied stress.

In practice, however, the compliance function for a particular component, or structure, may not be known, so that a different approach for calculating critical flaw sizes is necessary. One method which has been adopted is based on the stress analysis of an edge-cracked plate.

For a surface flaw, the critical defect size is given by:-

$$a_{cr} = K_{Ic}^2 \left[\frac{\phi^2 - 0.212(\sigma/\sigma_y)^2}{1.21\pi\sigma^2} \right] \qquad (20)$$

where K_{Ic} = plane strain fracture toughness

σ = gross working stress normal to the major axis of the flaw

σ_y = 0.2% proof stress

a_{cr} = critical depth of surface defect

ϕ = double elliptical integral

Let
$$Q = \left[\phi^2 - 0.212 \left(\sigma / \sigma_y \right)^2 \right].$$
(21)

where Q can be considered as the flaw shape parameter which allows for the geometry of the flaw.

Hence, from Equation 20 and Equation 21, we have for a *surface flaw* :

$$\left(\frac{a}{Q} \right)_{cr} = \frac{K_{Ic}^2}{1.21 \pi \sigma^2}$$
(22)

In the case of an internal flaw, the coefficient in the denominator of Equation 22 is taken as equal to unity.

Hence for an *embedded flaw* :

$$\left(\frac{a}{Q} \right)_{cr} = \frac{K_{Ic}^2}{\pi \sigma^2}$$
(23)

The relationship between the flaw shape parameter, Q, and aspect ratio, $a/2c$, is shown in Figure 3.6, for different σ/σ_y values. a is the semi-minor axis and $2c$ the major axis of an elliptical flaw.

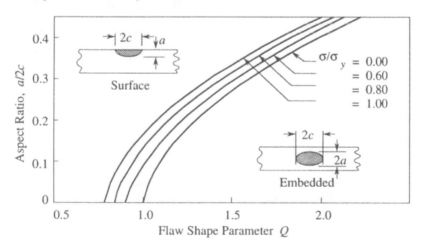

Figure 3.6 - Flaw Shape Parameter Curves for Surface Cracks

Values of Q for defects of various $a/2c$ ratios are given in Table 3.1.

Therefore, in any flaw size analysis problem it is necessary to have a knowledge of two of the following three quantities:

(i)   The fracture toughness (K_{Ic})

(ii)   The applied stress (σ)

(iii)   The flaw shape and size ($a/2c$)

In most cases, the flaw size will be the unknown quantity and before one can calculate the critical flaw size, it is necessary to assume the likely flaw aspect ratio ($a/2c$). Subsequent to this, ($a/Q)_{cr}$ can be determined for the specific value of σ and K_{Ic} involved. Hence, a and $2c$ can be derived.

This procedure can be reversed to calculate the gross failure stress for a given level of toughness and flaw size.

σ/σ_y	Q for $a/2c$ values of				
	0.10	0.20	0.25	0.30	0.40
1.0	0.88	1.07	1.21	1.38	1.76
0.9	0.91	1.12	1.24	1.41	1.79
0.8	0.95	1.15	1.27	1.45	1.83
0.7	0.98	1.17	1.31	1.48	1.87
0.6	1.02	1.22	1.35	1.52	1.90
<0.6	1.10	1.29	1.43	1.60	1.98

Table 3.1 - Flaw Shape Parameter Values

Example 3c - Calculation of Tolerable Size of Surface Flaw

During the water quenching of a steel component, 30 mm in section, heat transfer calculations indicate that the stress generated is 130 MPa. The K_{Ic} determined in laboratory tests was 30 MPa m$^{1/2}$ and the proof stress was 620 MPa.

The maximum size of surface defect specified in production is 0.50 mm.

(i) Calculate the tolerable defect size given that the aspect ratio of the flaw, $a/2c = 1/10$.

(ii) What would be the situation if the generated stress approached the proof strength of the material?

Solution

Case (i)

From Equation 22

$$\left(\frac{a}{Q}\right)_{cr} = \frac{K_{Ic}^2}{1.21\pi\sigma^2}$$

$$= \frac{30 \times 30}{1.21\pi 130 \times 130}$$

$$= 0.0140 \text{ m}$$

From Table 3.1, $Q = 1.1$ ($\sigma/\sigma_y = 0.21$, $a/2c = 0.1$)

$$\therefore a_{cr} = 15.4 \text{ mm}$$

Thus, the flaw would have to be greater than 15mm deep (i.e. halfway through the section) before fracture would occur. Clearly, there should be little risk of failure.

Case (ii)

From Equation 22,

$$\left(\frac{a}{Q}\right)_{cr} = \frac{30 \times 30}{1.21\pi 620 \times 620}$$

$$= 0.00062 \text{ m}$$

From Table 3.1, $Q = 0.88$ (since $\sigma/\sigma_y = 1$)

$$\therefore a_{cr} = 0.54 \text{ mm}$$

In this case, a flaw only 0.54 mm deep would be critical, which corresponds with the maximum depth specified in production. On this basis, it would be difficult from inspection to ensure that fracture did not occur upon quenching.

Example 3d - Calculation of Failure Stress for a Known Defect Size

Following the failure of a rocket motor casing during proof testing it was found that an internal flaw had extended to a size of 4 mm by 1.6 mm. The measured failure stress was 1260 MPa.

The material had been heat treated to a proof strength of 1645 MPa and had a K_{Ic} value of 60 MPa m$^{1/2}$.

Calculate the applied stress necessary to cause failure using the formula for an embedded flaw in Equation 23. Assume the ratio of σ/σ_y to be unity.

Solution

The aspect ratio, $\dfrac{a}{2c} = \dfrac{0.8}{4}$

(a_{cr} = *half* the width for an embedded flaw)

From Table 3.1, $Q = 1.07$ (since $\sigma/\sigma_y = 1$)

$\therefore$ From Equation 23,

$$\sigma = \sqrt{\frac{K_{Ic}^{2}}{\pi\left(\dfrac{a}{Q}\right)_{cr}}}$$

$$= \sqrt{\frac{60 \times 60 \times 1.07}{\pi 0.0008}}$$

$$= \mathbf{1238\ MPa}$$

The predicted stress of 1238 MPa shows close agreement with the observed failure stress of 1260 MPa.

3.5 Short Crack K_{Ic} Values

It has been stated (see B.S. 7448) that for a K_{Ic} value to be valid the fatigue pre-crack length must fulfil the criterion in equation 10 that

$$0.45 \leq a/W \leq 0.55.$$

This requirement applies to the fracture toughness test procedure. For values of K_{Ic} to have relevance to practical engineering assessments it is clearly important that values should either be strictly applicable to a range of crack lengths, or should demonstrably give conservative (i.e. safe) predictions. Strict applicability is more likely to hold for high strength, rather brittle materials. Figure 3.7 shows the fracture toughness values calculated from testpieces of G150 maraging steel of yield strength 2400 MPa for a range of a/W values. As long as the material has not exceeded any of the other criteria for fracture toughness evaluation, a crack length outside the Standard still gives the same result. This relies upon the use of a compliance function which is valid at the crack length in question. The function in B.S. 7448 holds between $0 \geq a/W \geq 1$, but results below $a/W = 0.1$ may become less accurate because the error in measuring the crack length is large with respect to the crack length.

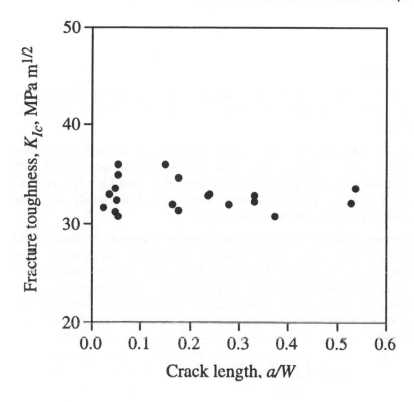

Figure 3.7

Note that these results would not hold to such low a/W values if the yield strength was substantially lower, because plasticity would invalidate the linear elastic assumptions of the stress analysis (see later).

4. Yielding Fracture Mechanics

4.1 Introduction

The principles of linear elastic fracture mechanics can be applied only in situations where the stress field is dominated by the stress intensity factor (see Section 2.5).

Examples treated so far have been concerned with high strength steels, alloy steels and aluminium alloys, where valid K_{Ic} results can be obtained with relatively small testpieces. However, it was seen, that even for a forging steel (Example 2j), quite large specimens were needed to satisfy the size requirements imposed by the British Standard.

This problem is accentuated when we consider materials of lower yield strength, such as structural steels at room temperature. For example, with A533B (0.23 C 1.5 Mn 0.5 Mo) steel, which is used to make welded pressure vessels in nuclear power reactors, the yield strength is approximately 500 MPa and a reasonable lower bound to the fracture toughness of fully stress-relieved plate at room temperature is 180 MPa m$^{1/2}$. Applying the standard size requirement:

$$B \geq 2.5 \left(\frac{K_{Ic}}{\sigma_y} \right)^2$$

gives B $\geq$ 0.324 m and W $\geq$ 0.648 m

This would equate to a weight of approximately 1500 kg in a CTS specimen and imply a fracture load of ~ 5 MN. Whilst the thickness is not unrepresentative of the service application (very large, thick vessels), the size of the testpiece needed to give a valid K_{Ic} value is clearly too large to be contemplated for routine purposes.

In the application of A533B, there is an obvious need to assess the effects of neutron irradiation on toughness. It is possible to subject specimens to an enhanced neutron flux by embedding them in the reactor core, but such surveillance specimens must of necessity be very much smaller than those needed for valid K_{Ic} tests. Other toughness parameters, which can be measured using small testpieces, must therefore be sought. An essential point to consider, therefore, is the effect of plasticity on the crack tip stress field.

4.2 Plasticity "Correction Factors"

An early approach treated the infinite body in plane stress, for situations where the plastic zone preceding fast fracture was considered as an additional contribution to the initial crack length. For a plastic zone of length $2r_y$, where r_y is given by:

$$r_y = \frac{a}{2}\left(\frac{\sigma_F}{\sigma_y}\right)^2 \tag{24}$$

it was argued that the distribution of stress was equivalent to that for an elastic crack of length $(a + r_y)$, i.e. :-

$$K = \sigma(\pi (a + r_y))^{1/2} \tag{25}$$

Although the energy relationship, $K^2 = EG$, is not proved for this definition of K, the expression is instructive in showing that, if the amount of plasticity is significant, the value of K developed by a given stress, σ, is higher than the elastic relationships indicate.

Example 4a - Effect of Local Crack Tip Plasticity

An aluminium alloy of yield stress 400 MPa, is tested in wide sheets. For a central crack of length 25.4 mm the fracture stress is observed to be 200 MPa. For a central crack of length 16.6 mm the fracture stress is 240 MPa.

Calculate values of fracture toughness for the alloy,

(a) using simple LEFM analyses

(b) applying the plasticity 'correction'.

Solution

Taking the stress analysis as that for an infinite body (see Table 2.1)

Case (a) - Elastic Solution

$$K = \sigma_{app}(\pi a)^{1/2}$$

For (i) $2a = 25.4$ mm $\therefore a = 0.0127$ m

$$K_c = 200 (0.0127\pi)^{1/2}$$

$$= 39.9 \text{ MPa m}^{1/2}$$

For (ii) $2a = 16.6$ mm $\therefore a = 0.0083$ m

$$K_c = 240 (0.0083\pi)^{1/2}$$

$$= 38.8 \text{ MPam}^{1/2}$$

Case (b) - With 'Plasticity Correction'

$$K = \sigma_{app}[\pi(a + r_y)]^{1/2}$$

At fracture

$$K_c = \sigma_F\left[\pi a\left(1 + \frac{\sigma_F^2}{2\sigma_y^2}\right)\right]^{1/2} \quad \text{(from Equation 24)}$$

For (i)

$$K_c = 39.9\,[1 + \tfrac{1}{2}(0.5)^2]^{1/2}$$

$$\therefore K_c = 42.3 \text{ MPa m}^{1/2}$$

For (ii)

$$K_c = 38.8\,[1 + \tfrac{1}{2}(0.6)^2]^{1/2}$$

$$\therefore K_c = 42.2 \text{ MPam}^{1/2}$$

Comments

The discrepancy between the elastic calculation and the 'corrected' result becomes greater as (σ_{app}/σ_y) increases, i.e. as the plastic zone size becomes larger.

As (σ_{app}/σ_y) is increased still further, this modification of the elastic stress analysis becomes increasingly inappropriate. It is also often the case that, although a large structure may fail before general yield, so that some modified stress analysis could be used, a much smaller test specimen of the same material would exhibit general yielding before fracture.

4.3 Crack Tip Opening Displacement

In circumstances of extensive plasticity, several toughness parameters have been proposed. One which has found particular favour in the UK has been the CRACK TIP OPENING DISPLACEMENT (CTOD) usually given the symbol δ.

In simple terms, CTOD is the opening of the crack at a position corresponding to the original (sharp) crack tip and a specific value of CTOD, δ_i, characterises the onset of crack extension. In some situations, total instability is coincident with δ_i, but, in others, such instability may not occur until a higher value, δ_c, is attained. This is associated with a finite amount of crack growth, Δa, so that:

$$\delta_c = \delta_i + \left(\frac{d\delta}{da} \right) \Delta a \tag{26}$$

In plane stress, for values of $\sigma_{app}/\sigma_y < 1$ (ie below general yield) it is possible to relate δ to the applied stress, σ_{app}, through the expression:

$$\delta = \frac{8}{\pi} \frac{\sigma_y}{E} a \ln \left[\sec \left(\frac{\pi \sigma_{app}}{2 \sigma_y} \right) \right] \tag{27}$$

For low values of σ_{app}/σ_y, the "ln sec" term may be expanded to the first order as follows:

$$\ln \left[\sec \left(\frac{\pi \sigma_{app}}{2 \sigma_y} \right) \right] = \ln \left[1 - \frac{\pi^2 \sigma_{app}^2}{8 \sigma_y^2} \right]^{-1}$$

$$= \ln \left[1 + \frac{\pi^2 \sigma_{app}^2}{8 \sigma_y^2} \right]$$

$$= \frac{\pi^2 \sigma_{app}^2}{8 \sigma_y^2}$$

Hence, for conditions of small scale yielding, Equation 27 becomes:

$$\delta = \frac{\sigma_{app}^2 \pi a}{\sigma_y E} \tag{28}$$

$$= \frac{K^2}{\sigma_y E}. \qquad \text{(see Equation 4, p7)}$$

or
$$\delta = \frac{G}{\sigma_y} \qquad \text{(see Equation 5, Example 2d, p10)}$$

By comparison with previous expressions , the toughness, G_c, is equated to $\sigma_y \delta_c$. Although the energetics of the situation for extensive plasticity do not predict instability, an analogy to the virtual work argument may be considered under conditions of small-scale yielding (cf Example 2d)

If
$$\delta \xi = \int_0^{\delta a} \sigma u dr,$$

where the stress σ remains constant at σ_y in the plane stress yielded region and the displacement is a constant at δ_c.

Hence, $$\delta\xi = \sigma_y\delta_c\delta a$$

Since $G = \dfrac{\delta\xi}{\delta a}$ per unit thickness

Then $$G_c = \sigma_y\delta_c \qquad\qquad (29)$$

The principle of CTOD testing is then that a critical value, δ_c, may be measured in a body (testpiece) which has undergone extensive yielding (or even general yielding), where linear elastic fracture mechanics cannot be applied, and this same value can then be used to calculate the toughness of a much larger body (structure) where σ_{app}/σ_y is small, so that the relationship $G_c = \sigma_y\delta_c$ can be employed.

For higher stress levels, δ_c *characterises* the fracture condition, but energy calculations show that the system of crack plus yielded zone is always in neutral equilibrium, so that a condition for instability cannot be derived.

4.4 CTOD Testing Details

The preferred testpiece is of SEN bend configuration, the same as that indicated in Figure 3.1, with thickness, B, equal to the thickness of material to be tested. The width, W, is equal to $2B$, and the specimen is precracked in fatigue, to obtain a/W values in the range 0.45 - 0.55. Testing is carried out in three-point bending, with a total loading span of $4W$. A subsidiary testpiece, with $W = B$, is also permitted.

The CTOD is obtained by converting a clip-gauge displacement, V_g, measured at knife-edges, situated at a height z above the top surface of the specimen, to a crack-tip value, δ. The fatigue crack length on the fractured surface is measured and an average value of a is obtained. The relationship between δ and clip-gauge displacement is given by:-

$$\delta = \frac{K^2\left(1-v^2\right)}{2\sigma_y E} + \frac{0.4(W-a))V_p}{0.4W+0.6a+z} \qquad\qquad (30)$$

where Vp is the plastic component of the displacement obtained by constructing a line from the point of interest on a load / displacement curve, parallel to the elastic loading line, and measuring the offset corresponding to zero load.

The first term in Equation 30 derives from a finite element calculation relating δ to K in plane strain (see Section 4.5) and the second term from the geometry of specimen deformation, assuming a centre of rotation at a distance 0.4 $(W-a)$ below the crack tip.

Example 4b - Calculation of CTOD from Clip Gauge Displacement

In a CTOD test made on a structural steel of yield strength 430 MPa, using a specimen of dimensions B = 25 mm, W = 50 mm, and pre-cracked to a depth of 26 mm, to what total crack tip displacement does a clip gauge plastic deformation of 0.33 mm correspond, if the load at this point is 50 kN, and the knife-edges are 2 mm thick? (Young's modulus is 200 GPa and Poisson's ratio is 0.3)

Solution
Elastic term (from Equation 30) is:

$$\delta = \frac{K^2\left(1-v^2\right)}{2\sigma_y E}$$

Where $K = \dfrac{PY_1}{BW^{1/2}}$ (see Equation 8)

For a/W = 26/50 = 0.52, Y_1 = 11.36 (see Figure 3.1)
Working in MN and m

$$K = \frac{0.05 \times 11.36}{0.025 \times (0.05)^{1/2}}$$
$$= 101.6 \text{ MPa m}^{1/2}$$

From above therefore:

$$\delta = \frac{(101.6)^2 (0.91)}{2 \times 430 \times 200 \times 10^3}$$
$$= 0.055 \text{ mm}$$

Geometrical term (Equation 23) is:

$$\delta = \frac{0.4(24) \times 0.33}{(0.4 \times 50) + (0.6 \times 26) + 2}$$
$$= 0.084 \text{ mm}$$

Thus the total (on-load) **CTOD = 0.14 mm**

The standard simply specifies the method by which critical values of CTOD, δ_c, may be obtained from clip gauge readings. The determination of δ_c is straightforward for a cleavage fracture, where fast fracture instability coincides with initiation, but, in other situations, it may be necessary to measure the initiation value, δ_i, or a value corresponding to a fixed amount of crack growth. Whilst fracture instabilities in testpieces can perhaps be related to fracture instabilities in service, it is hazardous to suppose that the onset of plastic instability in a testpiece (at δ_m) necessarily relates to any critical event in a structure.

It is possible, in some materials, to detect initiation of fibrous fracture, using acoustic emission or potential drop techniques, but these methods are not universally applicable. A technique which has proved useful in steels is to obtain a full curve of δ versus the crack extension Δa. Testpieces are unloaded from different positions on the load / clip-gauge displacement curve, heat tinted if necessary and fractured in liquid nitrogen. It is then possible to calculate the CTOD *at the position of the original crack tip* (taking into account any change in the centre of rotation due to crack growth) and to plot this against Δa. Extrapolation to $\Delta a = 0$, provides a value for δ_i.

Another technique is to unload specimens partially at various positions on the load / V_g displacement curve and use the elastic unloading slopes to detect any changes in compliance, which would indicate crack extension. A further method is to use two or more clip-gauges situated at different heights above the crack tip. The first major deviation from the initially linear proportionality of readings from gauges is indicative of crack initiation.

Example 4c - Determination of δ_i by Multi-Specimen Extrapolation Technique

The following table gives values of CTOD and fibrous thumbnail lengths for specimens of HY80 steel. What are the values of δ_i and $(d\delta/da)$ for this material?

Crack Extension (Δa)	CTOD (at original crack tip)
mm	mm
0.00	0.09
0.05	0.15
0.125	0.18
0.14	0.22
0.30	0.25
0.40	0.30
0.50	0.34
0.90	0.50

Table 4.1

4.5 CTOD and Fracture Toughness

In principle, a critical value of CTOD, measured in a testpiece, may be used to calculate the relationship between fracture stress and defect size in a structure. The first example employs a plane stress analysis.

Example 4e - Calculation of Critical Defect Size using CTOD

A rocket motor case is fabricated from Cr-Mo low alloy steel of proof strength 1.2 GPa, as a long cylinder of diameter 0.5 m and wall thickness 2.5 mm. If the design pressure at maximum thrust is 8 MPa, calculate the size of the largest defect that could be tolerated in the motor case without any risk of bursting at blast-off; given that the critical CTOD for the material is measured as 50 μm in a small testpiece.
(Young's modulus for the steel is 200 GPa.)

Solution
Assume that the conditions are those of plane stress, because the case is thin. The most dangerous defect will be a longitudinal crack, lying normal to the hoop stress.
The hoop stress is given by:

$$\sigma_h = p\,d\,/\,2\,t$$

where d is the diameter (0.5 m), t is the wall thickness (2.5 mm) and p is the pressure (8 MPa)

$$\sigma_h = \frac{8 \times 0.5}{2 \times 2.5 \times 10^{-3}} = 800 \text{ MPa}$$

$$\therefore K = 800\,(\pi a)^{1/2}$$

Now, in plane stress;

$$K_c^2 = EG_c = E\sigma_y\delta_c$$

$$= 200 \times 10^3 \times 1200 \times 50 \times 10^{-6}$$
$$= 12\,000$$

$$a_{crit} = \frac{K_c^2}{\sigma^2\pi}$$

$$= \frac{12000}{800 \times 800\pi}$$

$$= 5.8 \text{ mm}$$

The total length of defect that could be tolerated is therefore **11.6 mm**.

Comments
A rough estimate of the plastic zone radius is given by:

$$r_y = \frac{a}{2}\left(\frac{\sigma_F}{\sigma_y}\right)^2 = \frac{5.8}{2}\left(\frac{800}{1200}\right)^2$$

is 1.3 mm, so that $2r_y$ (2.6 mm) is about equal to the plate thickness. The plane stress assumption is therefore tenable.

In plane strain, no full analytical dependence of CTOD on applied stress has been produced, although some finite element solutions are becoming available. Under conditions of small 'scale' yielding, the results are of the form:

$$\delta = \beta\frac{K^2}{\sigma_y E} \tag{31}$$

where different calculations give values of β from 0.45 to 0.7, although some experiments have yielded a value of unity. Additionally, in structural steels, the appropriate critical value, δ_c, is not easy to deduce if slow ductile growth precedes the final fracture instability. A lower-bound estimate to K_{Ic} above the ductile / brittle transition temperature can be made from values of δ_i, but the value of K_i thus obtained may be quite substantially less than the K_{Ic} value in a large specimen, because the K_{Ic} Standard incorporates an amount of crack growth at P_Q. (See Section 3.3).

Example 4f - The Toughness of High-Strength Weld Metal

An alloy weld metal, of proof strength 1 GPa, was found to have a plane strain fracture toughness, K_{Ic}, of 95 MPa m$^{1/2}$. In CTOD tests, made on thick, small specimens, the value of δ_i was found to be 40 μm. What is the value of the constant, β, relating δ to $K^2/\sigma_y E$ in this material?
Young's modulus may be taken as 200 GPa.

Solution

$$\delta = \beta\frac{K^2}{\sigma_y E} \qquad \text{(Equation 31)}$$

Working in MN and m

$$\beta = \frac{0.04 \times 10^{-3} \times 10^3 \times 200 \times 10^3}{95 \times 95}$$

$$= 0.89$$

Hence, experimentally, for these plane strain tests,

$$\delta = 0.89\frac{K^2}{\sigma_y E}$$

Example 4g - Estimates of Toughness in Thick-Section Pressure Vessel Steel

Room temperature CTOD tests on stress-relieved A533B pressure-vessel steel gave a value for δ_i of 0.19 mm and a more-or-less linear increase of CTOD with crack extension (dδ/da) of 0.5 mm/mm. Calculate values of K_i for this material and estimate possible values of K_{Ic} that might be obtained in large LEFM testpieces. The 0.2% proof strength is 500 MPa , Young's modulus is 200 GPa and Poisson's ratio is 0.3.

<u>Solution</u>

The lowest value of the constant, β, relating to $K^2/\sigma_y E$ is given in the British Standard, where:

$$\beta = (1 - v^2)/2 \approx 0.45$$

Working in MN and m, from Equation 31,

$$K_i^2 = \frac{0.19 \times 10^{-3} \times 500 \times 200 \times 10^3}{0.45}$$

$$\therefore K_i = 206 \text{ MPa m}^{1/2}$$

The highest value of β, from experimental measurement, may be taken as unity.

Hence, from Equation 31,

$$K_i^2 = 0.19 \times 10^{-3} \times 500 \times 200 \times 10^3$$

$$\therefore K_i = 138 \text{ MPa m}^{1/2}$$

Comments

At room temperature, values of K_{Ic} measured experimentally were of the order 180 - 200 MPam$^{1/2}$, suggesting that the total fracture instability in this material occurs very shortly after initiation.

If a K value of 200 is taken, the 'valid' testpiece dimensions become:

$$B, a, (W - a) \geq 2.5\left(\frac{K}{\sigma_y}\right)^2$$

$$= 0.4 \text{ m}$$

The 5% offset secant construction (see Section 3.3) in the fracture toughness standard, permits an amount of plasticity and crack growth totalling 2% of the crack length. If the cracking is half this amount, it would be possible to contemplate as much as 4 mm slow crack growth preceding fast fracture at a temperature somewhat above room temperature. Such growth would imply a CTOD value:

$$\delta = \delta_i + \left(\frac{d\delta}{da}\right)\Delta a \qquad \text{(see Equation 26)}$$

$$= 0.19 + (0.5)\text{x}4$$

$$= 2.19 \text{ mm}$$

and $K = 690 \text{ MPa m}^{1/2}$ for $\beta = 0.45$

or $K = 460 \text{ MPa m}^{1/2}$ for $\beta = 1$

Similar effects have been noted, using J - integral calculations, and there is a real question as to whether a steel, such as stress relieved A533B, which has a high resistance to ductile growth, can show a fast fracture instability, in thick section, above its fibrous / cleavage transition temperature.

4.6 CTOD and Quality Control

Another use of CTOD is to provide a screening test for quality control purposes. Here, the aim is to specify the crack-tip ductility that a material must possess if it is not to fracture by cleavage in a plate of given thickness, B. The cause of such cleavage fracture is the high tensile stress that is generated ahead of the crack tip by triaxial constraint. If the plastic zone is sufficiently large, however, the triaxial stresses may be relieved by yielding on plates at 45° through the thickness.

The criterion suggested is that the plane-stress plastic zone radius, $r_y = \frac{1}{2}\pi$ $(K/\sigma_y)^2$, must be at least equal to the plate thickness, B. If δ, in plane stress, is given simply by:

$$\delta = \frac{K^2}{\sigma_y E} \qquad \text{(see Equation 31)}$$

substitution for K in terms of r_y gives:

$$\delta = 2\pi \, r_y \, (\sigma_y/E)$$

Writing the yield strain, ε_y, for (σ_y/E) and setting the condition that $r_y \geq B$ to avoid cleavage, we have, for quality control:

$$\delta \geq 2\pi \, \varepsilon_y \, B \qquad\qquad (32)$$

In practice, this value of CTOD may be equated to an equivalent minimum Charpy value.

Example 4h - Estimate of Minimum Crack-Tip Ductility

Estimate the minimum crack tip ductility (critical CTOD) required to prevent cleavage fracture in 25 mm thick structural steel plate at a temperature of 265 K, at which temperature its yield stress is 400 MPa. Young's modulus for the steel is 200 GPa.

Solution

The yield strain is given by:

$$\varepsilon_y = \frac{\sigma_y}{E}$$

$$= \frac{400}{200 \times 10^3}$$

$$\therefore \varepsilon_y = 2 \times 10^{-3}$$

From Equation 25,

$$\delta \geq 2\pi \times 2 \times 10^{-3} \times 25$$

$$= 0.314\text{mm}$$

Comment

Hence, if the steel possesses, at 265 K, in 25 mm thick testpieces, a CTOD greater than 0.314 mm, the plate should not fail by cleavage, whatever the initial defect length. If the CTOD is less than 0.314 mm, a more detailed fracture mechanics assessment must be carried out.

4.7 Defect Assessment (Welding Institute Method)

In lieu of detailed finite-element calculations, relating the applied load on a structure to the values of CTOD developed at defects situated in stress fields around major stress concentrators (such as a crack in a weld around a nozzle in a pressure vessel), the Welding Institute has developed a design procedure, using a curve, which relates CTOD to the local estimated strain level.

There are two components to the approach. Firstly, a wealth of data has been collected from tests on large pre-cracked plates, expressing crack tip CTOD as a function of applied strain level. Secondly, the strain concentration around a feature in service is equated to the stress concentration of the feature.

The design curve is shown in Figure 4.3 where the 'normalised' CTOD ($\delta / 2\pi \, \varepsilon_y a$) is plotted as a function of normalised strain ($\varepsilon/\varepsilon_y$).

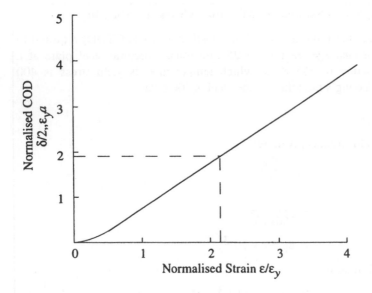

Figure 4.3

For a given value of $\varepsilon/\varepsilon_y$, the curve shows the value of δ/a that is developed. This information can be used in two ways: either to specify minimum values of δ for a given defect population, or to set NDT limits for material of known crack-tip ductility. Residual stress may also contribute to the value of $\varepsilon/\varepsilon_y$.

It should be noted that the value of a refers to the half length of maximum allowable through wall crack. The derivation of equivalent surface and embedded crack depths is explained in reference 14.

Example 4i - Use of Welding Institute Design Curve

A pressure vessel, of diameter 25m and wall thickness 0.25 m is made of low-alloy steel of proof strength 500 MPa and contains steam at a pressure of 5.5 MPa. Large pipes are welded into circular openings in the vessel wall using matching weld metal. If the available NDT techniques are able to detect defects of length 10 mm in the welds, calculate the minimum critical CTOD of the weld metal required to prevent fast fracture, assuming that a residual tensile stress equal to half the yield strength may be present in the region of the weld.

Solution

Since the diameter, d, is very much greater than the thickness, t, treat as a thin-walled tube:

Hoop stress
$$\sigma_h = \frac{pd}{2t}$$

$$= \frac{5.5 \times 25}{2 \times 0.25}$$

$$= 275 \text{ MPa}$$

Around the circular holes, the stress will be raised by a factor of x3, so that the proof stress is exceeded locally and the material is strained plastically.

If we assume that:

$$\frac{\sigma}{\sigma_y} = \frac{\varepsilon}{\varepsilon_y}$$

Hence,
$$\frac{\varepsilon}{\varepsilon_y} = \frac{825}{500}$$

$$= 1.65 \quad \text{from the external loading}$$

But
$$\frac{\varepsilon}{\varepsilon_y} = 0.50 \quad \text{from the residual stress}$$

Hence, Total
$$\frac{\varepsilon}{\varepsilon_y} = 2.15$$

From the Design Curve (see dotted line on Figure 4.3)

$$\frac{\delta}{2\pi\varepsilon_y a} = 1.9 \quad (\text{For } \frac{\varepsilon}{\varepsilon_y} = 2.15)$$

and

$$\varepsilon_y = \frac{500}{200 \times 10^3}$$

$$= 2.5 \times 10^{-3}$$

To be on the safe side, we can take a as an edge crack and substitute a value of 10 mm (i.e. NDT detection limit)

Hence, $\delta \geq 2\pi \times 2.5 \times 10^{-3} \times 10 \times 1.9$

$$\therefore \delta_c \geq 0.3 \text{ mm}$$

Comments

This value of δc should not be difficult to obtain in fully stress-relieved material above the cleavage / fibrous transition temperature but attention must be drawn to the importance of proper stress relief in welded regions. If residual stresses were to promote cleavage fracture in a brittle, untempered microstructure, the value of δ_c could be substantially reduced, to a value less than that required to give catastrophic propagation.

4.8 Defect Assessment ('Two Criteria' Approach)

This recognises that two distinct classes of failure can occur:
(a) Failure by fast fracture at an applied stress below the general yield stress (or collapse load).
(b) Failure by plastic collapse.
The existence of these two classes may be understood by considering the effect of decreasing the defect size in a large structure of constant toughness.
At large crack lengths, failure could occur under LEFM conditions, with the failure stress, σ_f, given by:

$$\sigma_f = K_{Ic} \, (\pi \, a)^{-1/2}$$

As a is decreased, so σ_f must increase, until it eventually becomes greater than the collapse stress σ_{GY}. Under these conditions, any fracture mechanics failure criterion is meaningless. Failure occurs by plastic collapse, although σ_{GY} must be calculated for cracked, rather than the uncracked body.

This simple concept is illustrated in normalised form in Figure 4.4 where the ordinate is the ratio L_f/L_{GY}, of the failure stress to the collapse stress and the abscissa is the ratio L_K/L_{GY} of the failure stress predicted by LEFM to the collapse stress.

As L/L_{GK}, $(\sigma_{app}/\sigma_{GY})$ becomes large, the linear elastic description of the stress field begins to break down. Use is then made of the 'ln sec' formula (Equation 27) to derive a 'true' value of K_C, from the plane stress expression:

$$\delta = \frac{K^2}{\sigma_y E}$$

Hence, $K_C = (E\,\sigma_y\,\delta_c)^{1/2}$

At high values of σ_{app}/σ_y, δ_c is given by:

$$\delta_c = \frac{8}{\pi}\frac{\sigma_y}{E}\,a\ln\sec\left[\frac{\pi\sigma_F}{2\sigma_y}\right]$$

Substituting, $K_C = (E\,\sigma_y\delta_c)^{1/2}$, $K_Q = \sigma_F\,(\pi a)^{1/2}$

$$K_c^2 = \frac{8}{\pi}\sigma_y^2 a\ln\left[\sec\left(\frac{\pi\sigma_F}{2\sigma_y}\right)\right]$$

or $K_c^2 = \frac{8}{\pi}\sigma_y^2 a\ln\left[\sec\left(\frac{K_Q}{2\sigma_y}\right)\left(\frac{\pi}{a}\right)^{1/2}\right]$ (33)

which is one form of the expression used to deduce 'true' K_C values from 'apparent' K_Q values.

In terms of the normalised stresses, L_f, L_K and L_{GY}, write:

$$K_Q = L_f(\pi a)^{1/2}$$

$$K_C = L_K(\pi a)^{1/2}$$

$$L_{GY} = \sigma_y$$

Hence, the equation becomes:

$$L_K^2 \pi a = \frac{8}{\pi}L_{GY}^2 a\ln\left[\sec\left(\frac{\pi L_f}{2 L_{GY}}\right)\right]$$ (34)

which gives (L_f/L_{GY}) as a function of (L_K/L_{GY}), plotted as the failure curve in Figure 4.4.

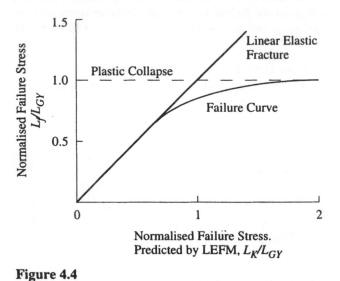

Figure 4.4

4.9 The R6 Approach

In yet another formulation, due to Milne et al, (ref. 17) the ratio L_f/L_K is written as K_r^f and L_f/L_{YG} is written as S_r^f.

From Equation 34,

$$\left(\frac{L_K}{L_f}\right)^2 = \frac{8}{\pi^2}\left(\frac{L_{GY}}{L_f}\right)^2 \ln\left[\sec\left(\frac{\pi L_f}{2L_{GY}}\right)\right]$$

$$\frac{1}{K_r^f} = \frac{1}{S_r^f}\left[\frac{8}{\pi^2}\ln\left[\sec\left(\frac{\pi L_f}{2L_{GY}}\right)\right]\right]^{1/2}$$

$$K_r^f = S_r^f\left[\frac{8}{\pi^2}\ln\left[\sec\left(\frac{\pi}{2}S_r^f\right)\right]\right]^{-1/2} \qquad (35)$$

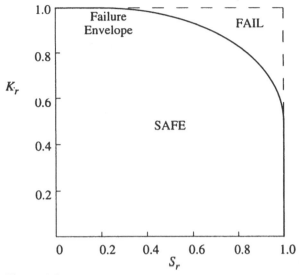

Figure 4.5

and this expression is plotted as the 'failure assessment' curve Figure 4.5, although this may incorporate a 'collapse load', L_u, instead of L_{GY}, where L_u is calculated using the same constraint using the same constraint factor, but with σ_y replaced by the tensile strength σ_u, or an average flow stress, equal to $\frac{1}{2}(\sigma_y + \sigma_u)$. Such expressions are held to provide a useful lower bound to experimental results obtained on specimens which fracture at fairly high fractions of, and even beyond, general yield.

This approach may draw conclusions, as to the danger of defects, rather different from those using the Welding Institute method (previous Section 4.6). As for the Welding Institute method, it suffers from having a basis which is a mixture of plane stress analysis and empiricism. This approach is perhaps more open to question in situations where a *small test specimen* breaks *well after* general yield, but an actual structure made from the same material could fail before general yield.

If plastic collapse of the test specimen has occurred, it is not clear what parameter can be used in design.

The modified 'collapse' based on σ_u is empirical and σ_{app}/σ_u values in a generally yielded testpiece do not have obvious meaning with respect to the prediction of failure in a structure.

Example 4j - Stress Analysis When Yielding is Extensive

Show formally that expansion of Equation 27 to second order terms leads to an expression for K similar to that used in Example 4a. For the aluminium alloy described there, what critical defect size corresponds to a failure stress of 360 MPa? The yield strength of the material is 400 MPa.

Solution
Referring to Equation 27,

$$\delta_c = \frac{K_c^2}{\sigma_y E} = \frac{8}{\pi} \frac{\sigma_y}{E} a \ln \left[\sec \left(\frac{\pi \sigma_F}{2\sigma_y} \right) \right]$$

Let $\left(\dfrac{\pi \sigma_F}{2\sigma_y} \right) = \theta$ and expand to second order terms

Hence,

$$\ln \sec \theta = - \ln \cos \theta$$

$$= -\ln \left(1 - \frac{\theta^2}{2!} + \frac{\theta^4}{4!} + ... \right)$$

$$= -\ln \left[\left(1 - \left(\frac{1}{4} + \frac{\sqrt{12}}{24} \right) \theta^2 \right) \left(1 - \left(\frac{1}{4} - \frac{\sqrt{12}}{24} \right) \theta^2 \right) \right]$$

From which it can be shown, to second order terms, that

$$\ln \sec \theta = \frac{1}{2} \left(\theta^2 + \frac{\theta^4}{6} \right)$$

Hence Equation 27 becomes

$$\delta_c = \frac{K_c^2}{\sigma_y E} = \frac{8}{\pi} \frac{\sigma_y}{E} a \left[\frac{1}{2} \left(\frac{\pi^2 \sigma_F^2}{4\sigma_y^2} + \frac{\pi^4 \sigma_F^4}{6x16\sigma_y^4} \right) \right]$$

$$\therefore K_c^2 = \frac{4}{\pi} \sigma_y^2 a \left[\frac{\pi^2 \sigma_F^2}{4\sigma_y^2} + \frac{\pi^4 \sigma_F^4}{96\sigma_y^4} \right]$$

$$= \frac{\sigma_F^2 a}{\pi} \left[\pi^2 + \frac{\pi^4 \sigma_F^2}{24\sigma_y^2} \right]$$

$$= \sigma_F^2 \pi a \left[1 + \frac{\pi^2}{24} \frac{\sigma_F^2}{\sigma_y^2} \right]$$

$$K_c = \sigma_F (\pi a)^{1/2} \left[1 + \frac{\pi^2}{24}\frac{\sigma_F^2}{\sigma_y^2}\right]^{1/2}$$

$$K_c = \sigma_F (\pi a)^{1/2} \left[1 + 0.41\frac{\sigma_F^2}{\sigma_y^2}\right]^{1/2}$$

which should be compared with the following expression used in Example 4a:-

$$K_c = \sigma_F (\pi a)^{1/2} \left[1 + 0.5\frac{\sigma_F^2}{\sigma_y^2}\right]^{1/2}$$

For $\sigma_{app}/\sigma_y = 0.5$, the correction factors applied to the $\sigma_F(\pi a)^{1/2}$ term are 1.05 and 1.06 respectively.

For $\sigma_{app}/\sigma_y = 0.6$, the factors are 1.07 and 1.086.

Taking
$$K_c^2 = \frac{8}{\pi}\sigma_y^2 a \ln\left[\sec\left(\frac{\pi\sigma_F}{2\sigma_y}\right)\right] \quad \text{(ie Equation 27)}$$

$$K_c^2 = \frac{8}{\pi^2}\left(\frac{\sigma_y}{\sigma_F}\right)^2 \ln\left[\sec\left(\frac{\pi\sigma_F}{2\sigma_y}\right)\right]\sigma_F^2 \pi a$$

with $\sigma_F/\sigma_y = 0.5$ and 0.6 gives factors applied to $\sigma_F(\pi a)^{1/2}$ of 1.06 and 1.094. Agreement appears to be better with the simple plastic zone correction than with the expansion to second-order terms. The corrected values for K_c in Example 4a using the `ln sec' formula become 42.3 and 42.4 MPa m$^{1/2}$.

If the fracture stress is 360 MPa and the yield stress is 400MPa, we can write

$$K_c^2 = \frac{8}{\pi}\sigma_y^2 a \ln\left[\sec\left(\frac{\pi\sigma_F}{2\sigma_y}\right)\right]$$

as
$$(42.35)^2 = \frac{8}{\pi}(400)^2 a\ln[\sec 81°]$$

$$a = 2.37 \text{ mm}$$
$$2a = 4.7 \text{ mm}$$

Comments

Hence, a defect of length 4.7 mm would cause failure at an applied stress of 360 MPa.

Note that the LEFM treatment for this stress and crack length would give

$K_Q = 360 (0.00237\pi)^{1/2} = 30.9$ MPa m$^{1/2}$ only.

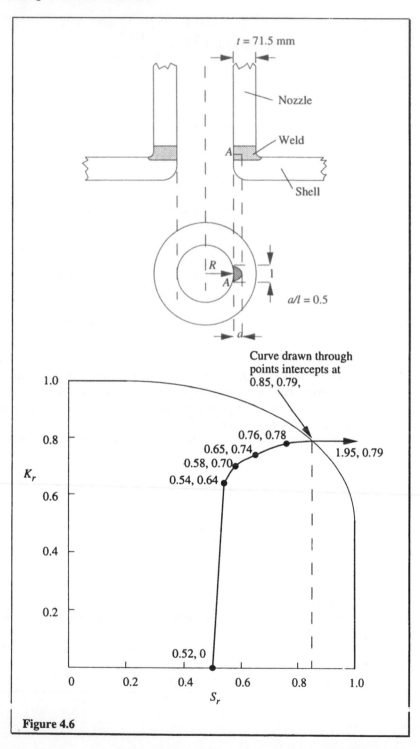

Figure 4.6

Example 4k - Failure Assessment using the R6 Approach

A set-on nozzle is welded to a pressure vessel shell, as indicated in Figure 4.6.
It is assumed that a thumbnail crack of aspect ratio $a/l = 0.5$ and a/t variable, may be present at location A. Using the information in table 4.2, calculate the critical size of this crack needed to give catastrophic failure, if the membrane stress is a simple hoop tension σ_h, of 210 MPa. The materials have the following properties: 0.2% proof stress = 420 MPa; tensile strength = 540 MPa; K_{Ic} weld metal = 80 MPa $m^{1/2}$

a, mm	a/t	S_r	K_r	$f(a/R)$
0	0.00	0.52	0.00	3.00
10	0.14	0.54	0.64	1.80
20	0.28	0.58	0.70	1.40
30	0.42	0.65	0.74	1.20
40	0.56	0.76	0.78	1.01
50	0.70	1.95	0.79	1.00

Table 4.2

Solution
Successive pairs of values of S_r and K_r corresponding to different values of a/t (eg 0.52, 0; 0.54, 0.64; 0.58,0.70; etc) are plotted as points on the failure assessment diagram (see Figure 4.6).
The line joining these points crosses the failure envelope at
$S_r = 0.85$ and $K_r = 0.79$
i.e. when a = 45 mm
Failure is predicted for crack depths **greater than 45 mm.**

Comments
The values of S_r have been calculated by finite element analysis, using the membrane stress of 210 MPa, and flow stress equal to $\frac{1}{2}$(proof stress + tensile strength), allowing for the change in ligament area as a/t varies. Values of K are given by:
$$K = \sigma_h (\pi a)^{1/2} f(a/R)$$

where $f\ (^a/_R)$ is a factor to take account of the stress gradient beyond the circular nozzle.

When $^a/_R$ is zero, f = 3.

When $^a/_R$ is large, f = 1.

For a = 45 mm, the table 4.2 shows that $f \approx 1$.

It is interesting to note that the simple expression $K_{Ic} = \sigma_h\ (\pi a)^{1/2}$, which does not incorporate the factor of 1.12 for an edge crack (see Table 2.1), nor any factor for elipticity of the thumbnail profile (see Section 3.4), gives, with $K_{Ic} = 80$ MPa m$^{1/2}$, $\sigma_h = 210$ MPa and $f = 1$,

$$a = \frac{1}{\pi}\left(\frac{80}{210}\right)^2 \times 10^3$$

$$= 46 \text{ mm}$$

The R6 Assessment has now been modified to take account of materials which have a high initial rate of strain hardening. The S_r axis of the failure assessment diagram (figure 4.5) has been replaced by L_r which is the applied stress divided by the yield stress. The modified failure assessment diagram, figure 4.7, has cut-offs in L_r corresponding to specific flow stress / yield stress values appropriate to various classes of material with different strain hardening characteristics.

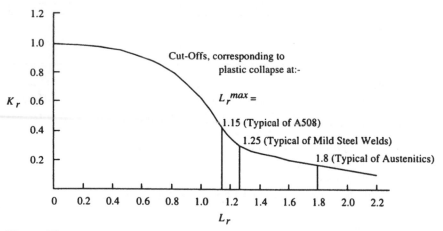

Figure 4.7

This revised diagram forms the basis of the assessment diagram for PD 6493:1991. A very convenient way to assess defects in structures following PD 6493:1991 procedures is to use interactive software packages, such as 'CRACKWISE, available from TWI, Abington Hall, nr. Abington, Cambridge.

4.10 Non-Linear Elastic Behaviour

For a material which behaves in a linear elastic manner (the graph of stress vs. strain is a straight line). However there are types of material e.g. rubbers, which behave in a non-linear elastic manner such that the stress-strain graph is a curve. The material still behaves elastically in that loading, unloading and re-loading follow the same curve.

The energy release rate associated with notional crack extension in non-linear elastic materials is characterised by a parameter termed J, which is the non-linear equivalent of the potential energy release rate, G per unit thickness derived for linear elastic materials (section 2.3). J is defined as follows

$$J = - dU_{tot}/Bda \qquad (36)$$

where U_{tot} is the total potential energy within the system. J is the non-linear elastic equivalent to G, and in a linear elastic material J would be identical to G. For a testpiece containing a crack and subjected to stress, the potential energy is composed of two parts.

$$U_{tot} = U_0 + U_a \qquad (37)$$

where U_0 is the potential energy associated with the testing system loading an un-notched testpiece and U_a is a component arising from the fact that the testpiece contains a crack of length a. If a plot of load against load-line-displacement is made, this second component, U_a, is the area under the trace. This can be expressed as

$$U_a = \int P(a) \, dq \qquad (38)$$

where q is the load-line-displacement. If the crack is extended by a small amount Δa, under fixed displacement (i.e. keeping q constant) the load decreases and the change in total energy is

$$\Delta U_{tot} = U_0 + \int P(a) \, dq - U_0 - \int P(a + \Delta a) \, dq \qquad (39)$$

$$= \int \Delta P \, dq \qquad (40)$$

for infinitesimal crack extension

$$dU_{tot} = \int dP \, dq \qquad (41)$$

$$J = -\int \left(\frac{dP}{da}\right) dq \qquad (42)$$

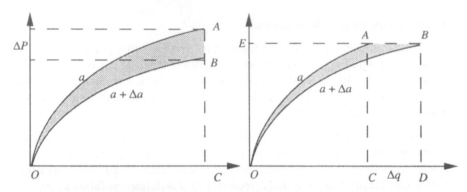

Figure 4.8 - Crack extension for constant displacement conditions

Figure 4.9 - Crack extension for constant load conditions

This can be seen graphically in fig. 4.8. The crack of length $a+\Delta a$ can be seen to give a lower load for a constant displacement. As the stored strain energy, U_a, in the testpiece is the area under the curve, then for the crack of length a this is *OAC*. For the increased crack length this becomes *OBC* and the release of stored strain energy is *OAC-OBC*. As

$$J = \frac{\Delta U}{\Delta a}$$

Then $\quad\quad\quad J\Delta a = \text{area } OAB = \int_0^q \Delta P dq$

as $\Delta a \to 0$ then $\quad J = \int_0^q \left(\frac{dP}{da}\right) dq$

If we now take the case where the load is held constant then the displacement q will change (Fig. 4.9). In this case the work done $= P\Delta q = ABCD$ and is negative as this is work obtained. The *increase* in strain energy $= OBD - OAC$

as $\quad\quad\quad OBD = ODBE - OBE$

and $\quad\quad\quad OAC = OCAE - OAE$

then $\quad\quad\quad OBD - OAC = ODBE - OCAE - (OBE - OAE).$

As $\quad\quad\quad ODBE - OCAE = ABCD$

then $\quad\quad\quad ABCD = P\Delta q - OAB.$

The *net* energy release $= -P\Delta q + (P\Delta q - OAB) = -OAB.$

and the equation becomes

$$J\Delta a = \text{area } OAB = \int_0^P \Delta q dP \quad\quad\quad (43)$$

and then *J* becomes

$$J = \int_0^P \left(\frac{dq}{da} \right) dP \tag{44}$$

Another way of looking at this equation for J is that the axes of figure 4.9 are swapped round such that the load is along the x-axis and the displacement is along the y-axis.

It is, then, possible to provide J-calibrations for specific testpiece geometries by measuring the area OAB for fixed displacement or constant load as appropriate. In principle this may have to be measured using a planimeter (or counting squares on the graph), but analytical techniques are also available (see later). The critical value of J at fracture initiation (often defined as a crack extension of 0.2 mm) may be regarded as a fracture toughness, equivalent to G_{crit}. In the following example, we choose an analytical non-linear (parabolic) relationship between P and q to illustrate the general principle.

Example 41- Calculation of J for a Non-Linear Elastic Material

The graph of load-against-load-line displacement for a non-linear elastic material is characterised by the equation

$$P = 3.16 \times 10^5 \, (1 - a/W) \, q^{1/2}$$

where P is the load in N, and q is the load point displacement in m. The testpiece dimensions are $W = 10$ mm, $B = 5$ mm and the initial crack length of 5.0 mm grows to 5.2 mm. Calculate the J value if

(i) the load point displacement remains constant at 1 mm,

(ii) the load remains constant at 5 kN.

Solution

(i) From $dU_{tot} = \int \Delta P \, dq$ and $P = 3.16 \times 10^5 \, (1 - a/W) \, q^{1/2}$

we have $dU_{tot} = \int \Delta(3.16 \times 10^5 \, (1 - a/W) \, q^{1/2}) \, dq$

$dU_{tot} = 3.16 \times 10^5 \, [1 - 0.52 - 1 + 0.5] \, 2 \, q^{3/2} /3$

$= - 4216.4 \times 0.001^{3/2}$

Now $\qquad J = \int - \Delta U / B \Delta a$ (remember J is defined as being per unit thickness)

$\therefore \qquad J = 0.133.3 / (0.0002 \times 0.005)$

$J = 133.3 \text{ kJ m}^{-2}$

(ii) $dU_{tot} = -\int \Delta q \, dP$

If q is now expressed as a function of P, we have effectively swapped the load and displacement axes over

Now $q = \dfrac{P^2}{10^{11}(1-a/W)^2}$

$$dU_{tot} = -\int \Delta \left(\frac{P^2}{10^{11}(1-a/W)^2} \right) dP$$

$$= -10^{-11} P^3 \left(\frac{1}{0.48^2} - \frac{1}{0.5^2} \right)\frac{1}{3} = -10^{-11} P^3 \,(1/0.5^2 - 1/0.6^2)\,/\,3$$

$$= -0.1417$$

Now $J = -dU/Bda$ (remember J is defined as being per unit thickness)

∴ $J = 0.1417 / (0.0002 \times 0.005)$

$J = 141.7$ kJ m^{-2}

Comments

There is a significant difference between the values obtained by holding the load point displacement or the load constant. This difference is reduced if the crack growth is smaller. For crack growth of 0.01 mm (from 5 mm to 5.01 mm) the J value for constant load point displacement remains at 133.3 kJ m^{-2} whereas the J value for constant load becomes 133.6 kJm^{-2}. As for linear elastic behaviour, the values approach each other as $\Delta a \to 0$. This method may be obviously be extended for any power-law relationship between P and q (this reflects a power law dependency of stress on strain e.g. $\sigma = K\varepsilon^n$.)

4.11 J as a Characterising Parameter

As defined above, J is a non-linear elastic energy release rate which becomes identical to G for linear elastic behaviour. It will be recalled that G can also be related to the intensity of the stress and strain fields around a stressed crack tip through the identity (calculated from a virtual work argument) $G = \alpha K^2 / E$, (Example 2d pp9-10) where K is the (linear elastic) stress intensity factor. For material whose stress-strain curve is represented by a power law relationship between stress and strain ('power law hardening' material) the stresses do not follow the $r^{-1/2}$ relationship which holds for linear elastic behaviour, but are distributed in a manner such that the exponent of r is a function of the work hardening exponent. Thus, if the basic stress strain relationship is of the form $\sigma \propto \varepsilon^n$.

the stress field is given by $\sigma \propto \dfrac{\sigma_{app}}{r^{\frac{1}{n+1}}}$

and the strain field is $\varepsilon \propto \dfrac{\varepsilon_{app}}{r^{\frac{1}{n+1}}}$

The displacement field in the open crack is of the form $u \propto u_0 \, r^{\frac{1}{n+1}}$ and the justification for denoting the strengths of fields by J is that the virtual work argument for notional crack extension remains as $J = \int_a^{a+\Delta a} \sigma \, du$ as for the G / K relationship.

4.12 Path Independence of J

We have so far treated evaluations of J for remote fixed-grip (load point displacement) conditions, for remote constant load, and, via the virtual work argument, for the stress / displacement field at the crack tip. These treatments are entirely analogous to those used for G and the point to be made is that, for a given applied stress, the crack tip value of J (calculated from stress analysis) is the same as that determined from remote boundary conditions (which, in principle, can be determined experimentally). The value of J is the same for two circuits or 'paths':- one from the lower surface of the crack tip region round to the upper surface: the other following all external surfaces of the (2D plane section of the) testpiece. Both paths encompass the crack tip (singularity) and the same value of J is obtained. It can be shown theoretically that for any continuous path within the body, starting from the lower crack surface, encompassing the crack tip (singularity) and ending on the top crack surface, the value of J remains constant. The statement is made that J is a 'path-independent integral'. The use of the word 'integral' simply implies that J is evaluated within and all the way around the chosen path. For a given path J is defined as the difference between two terms: the first, the integrated strain-energy density within the path; the second, the work done by the surface tractions which would exist along the path's boundary if it were cut from the strained body without change of shape, moving through the appropriate displacements to remove these tractions to zero. In the limit of load controlled J-calibration, the first term is equivalent to the increase in strain energy stored

($P\Delta q$ - OAB, fig. 4.9):the second to the work done by the applied load ($P\Delta q$, fig 4.9).

4.13 Elastic/Plastic Behaviour

Although the *J*-Integral is defined strictly for non-linear elastic behaviour it has been applied to materials which behave in an elastic / plastic manner. In an elastic / plastic material the graph of load versus load-line-displacement curve is different on unloading from that on loading, because permanent plastic work remains stored in the specimen. It is apparent that *J*, defined as a potential energy release rate, is inapplicable to this situation. It is however, possible to define a J-like parameter '*J*' in terms of the *work input* rate

$$\text{'}J\text{'} = dU/Bda.$$

This '*J*' parameter characterises the work-rate put into a cracked testpiece, but this work is *not* fully recoverable to drive a crack. In Eshelby's words 'it is all right, so long as we do not call the material's bluff by unloading'. It is, however, plausible and open to experimental test, that, if a given value of '*J*' characterises a given amount of crack extension in one geometry, the same value of '*J*' will characterise the same amount in another geometry. It is, therefore, possible that a critical value of '*J*' measured on a small testpiece and corresponding to 0.2 mm of growth for example, would characterise 0.2 mm in a cracked structure. The experimental procedures require a means of measuring values of '*J*' in a testpiece. It has become conventional to omit the quotation marks and refer to '*J*' as *J*. This practice is now adopted, but it must be remembered that *J* for an elastic/plastic material is a *characterising parameter* and does not bear the full connotation of energy release. Elastic / Plastic finite element analysis shows that this new *J* is sensibly path-independent for a number of standard geometrics.

For elastic / plastic material, *J* is defined as the work input rate during the loading of a testpiece. This work rate can be used to produce either elastic or plastic deformation. Equations 42 and 44 still hold but they are modified to enable *J* to be determined from a single test.

If it is assumed that deformation only occurs in the ligament *b* (= *W-a*) it follows that d/da = -d/db. As *J* is measured per unit thickness it is necessary to express the load *P* as *P** = *P/B*. The displacement, q, is a function of ($P*/b^2$) and various material constants governing the stiffness. Hence $q = f(P*/b^2)$ and as the amount of deflection due to the un-notched testpiece is small, *q* can be assumed to be due almost entirely to the presence of the crack. Since d/da = -d/db

$$\frac{d}{da}q = -\frac{d}{db}q \tag{45}$$

$$= -\frac{d}{db}f\left(\frac{P*}{b^2}\right) \tag{46}$$

If this is written as $-\dfrac{d}{db}f(t)$ where $t = P*/b^2$, we can use the relation $\dfrac{d}{db} = \dfrac{d}{dt}\dfrac{dt}{db}$ to obtain

$$\frac{d}{da}q = 2P*f'\left(\frac{P*}{b^2}\right)\frac{1}{b^3} \tag{47}$$

where $f'(t)$ implies df/dt. Now $f(t)$ is a function of both $P*$ and b we can also write $\dfrac{d}{dP*} = \dfrac{d}{dt}\dfrac{dt}{dP*}$ and hence

$$\frac{d}{dP*}f\left(\frac{P*}{b^2}\right) = f'\left(\frac{P*}{b^2}\right)\frac{1}{b^2} \tag{48}$$

then substitute $f(P*/b^2)$ to obtain

$$\frac{dq}{da} = \frac{2P'}{b}\left(\frac{dq}{dP'}\right) \tag{49}$$

Substituting this into Equation 44 and replacing $P*$ by P/B, for crack extension, keeping displacement constant,

$$J = \int_0^P \frac{2P}{Bb}\left(\frac{dq}{dP}\right)dP \tag{50}$$

Hence

$$J = \frac{2}{B(W-a)}\int_0^q P\, dq \tag{51}$$

where $B(W-a)$ is the cross section of the unbroken ligament. It therefore becomes straightforward to obtain a J value from a test. It can be seen from Equation 38 that the integral can be replaced by the area under the load / load-line-displacement curve. Hence Equation 51 becomes

$$J = \frac{2}{B(W-a)}U_i \tag{52}$$

where U_i is the area under the curve up to the displacement of interest. A line is drawn parallel to the elastic loading line to pass through the curve at this point.

The area, U_p between the line that has been drawn, the loading trace and the abscissa is calculated, and J is then given by

$$J = \frac{K^2}{E}\left(1-v^2\right) + \frac{2U_p}{B(W-a)}$$

(53)

where $K = P Y_1 / (B \sqrt{W})$.

4.14 J Testing Details

The British Standard (BS 7448) describes two testpiece geometries, the SEN bend testpiece and the CTS testpiece. The CTS testpiece is stepped at the top of the notch to allow for the positioning of clip gauges (figure 4.10). The SEN bend testpiece, as shown in figure 3.1, has $B = 2W$ although a testpiece geometry of $B = W$ is also permitted and testing is carried out in three point bending with a total span of 4W. Both types of testpiece are fatigue precracked to $a \geq 0.45W$.

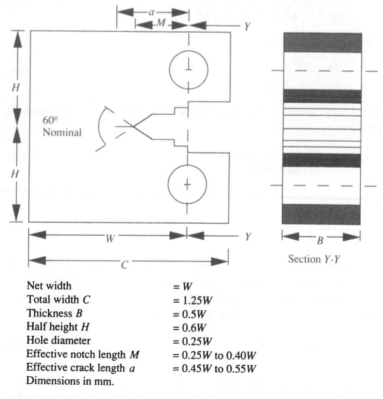

Net width	$= W$
Total width C	$= 1.25W$
Thickness B	$= 0.5W$
Half height H	$= 0.6W$
Hole diameter	$= 0.25W$
Effective notch length M	$= 0.25W$ to $0.40W$
Effective crack length a	$= 0.45W$ to $0.55W$
Dimensions in mm.	

Figure 4.10

During the test a trace of load against load line displacement needs to be obtained. The true load line displacement may be measured in a number of ways:

(i) using a comparator bar.

(ii) placing an un-notched testpiece in the test machine, with the rollers close together, measuring the elastic response of the system and subtracting this from the load crosshead displacement trace.

(iii) Using a clip-gauge arrangement to monitor the crack opening displacement and relating this to the load-line-displacement.

After the test the area U_p must be measured and this can be done either directly from the test record or by numerical integration using computer techniques. The value of J can then be obtained from equation 53.

Example 4m - Calculation of J for a SEN bend Testpiece.

A testpiece of NiCrMo steel of dimensions $B = 10$ mm, $W = 20$ mm, gave the load / load-line-displacement trace shown in Figure 4.11. If the crack length is 9.83 mm calculate the value of J for the material. The area U_p was found to be $4.820\,\text{Nm}^2$. Young's modulus may be taken as 206 GPa and Poisson's ratio as 0.3.

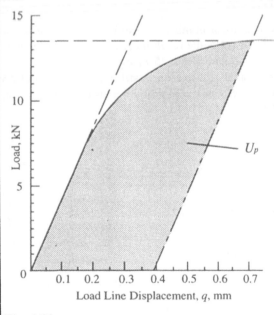

Fig. 4.11

<u>Solution</u>

The value of the maximum load is 13.5 kN.
Now

$$J = \frac{K^2}{E}\left(1 - v^2\right) + \frac{2U_p}{B(W - a)}$$

$B = 0.01$ m, $W = 0.02$ m, $a = 0.00983$ m, $W - a = 0.01017$ m,

$\therefore a/W = 0.4915$ and $Y_1 = 10.38$ (from figure 3.1)

$$K = \frac{PY_1}{B\sqrt{W}} = \frac{13.5 \times 10^3 \times 10.38}{0.01 \times \sqrt{0.02}} = 99.09 \, MPa \, m^{1/2}$$

$$\therefore \quad J = \frac{\left(99.09 \times 10^6\right)^2 (0.91)}{206 \times 10^9} + \frac{2 \times 4.820}{0.01 \times 0.01017}$$

$$= 43372 + 94789$$

$$J = 138.2 \text{ kJ m}^{-2}$$

The J determination of initiation (0.15 mm crack growth) can be carried out using two other methods, one similar to the method for evaluating CTOD in Example 4c and the other using an unloading technique.

4.15 Evaluation of J_{Ic}.

Having obtained J values for a number of testpieces it is possible to obtain a value for J_{Ic}. Firstly, to be valid, the result must obey the criteria that both

$$B \text{ and } W - a > 25 \, J / \sigma_0.$$

However, results which satisfy these criteria still may not be valid J values. To determine the validity of the results a plot of J versus crack extension, Δa, must be plotted (figure 4.12).

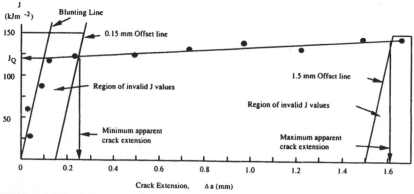

Figure 4.12 - The plot of J versus crack extension

Then a number of constructions must be added. First, the blunting line,

$$J = \left(\sigma_{ys} + \sigma_{UTS}\right)\Delta a,$$

must be drawn, followed by two lines parallel to the blunting line but with offsets of 0.15 mm and 1.5 mm. These two offset lines determine the limits of valid J values. A best fit line through the valid J values (called the R line) is then extrapolated to the blunting line to give the J_Q value for the material. The points where the R line cross the two offset lines are the minimum and maximum apparent crack extension. The J_Q value may be called the J_{Ic} value if there are at least four valid J values and the first valid value is at a crack extension of less than 0.6mm.

4.16 Unloading Compliance J_{Ic} test

During a normal fracture mechanics type test where load and CTOD are monitored, the testpiece is unloaded approximately 10% a number of times during the test. From these unloading lines an estimate of the crack growth may be obtained at each point on the curve and a point on the J versus Δa plot is obtained. This plot may be evaluated, in the manner described above, to obtain a value of J_{Ic}.

This method is a popular test procedure as it only requires a single testpiece. The experimental procedure is ideally suited to computerisation and this makes the test conducive for quality control testing.

4.17 Defect Assessment Using J

Values of J_{Ic} may be used in defect assessment procedures (e.g. PD 6493:1991) by calculating equivalent values of the critical stress intensity through the expression

$$K^2 = \frac{E J}{\alpha}$$

This is clearly analogous to equations 5 and 6 for linear elastic behaviour, with G replaced by J. Such values of K are often referred to by the symbol K_J. The ordinate of the assessment diagram (figure 4.9) is then expressed as $K_r = K_{app} / K_J$ and assessment proceeds as described in section 4.7. Detailed assessments are carried out using computer assisted methods (c.f. the "CRACKWISE' package mentioned at the end of section 4.9).

5. Fatigue Crack Growth

If a material is subjected to a cyclic stress which is well below the yield strength mechanical failure would not generally be expected. However, if there is a feature in the component such as a crack, an inclusion or a sharp geometrical discontinuity the stress may be raised locally to a level greater than the yield strength and crack initiation and growth may occur. As the applied stresses are cycled the crack may extend by a small amount on each cycle. This is known as fatigue crack growth.

5.1 The Paris-Erdogan Equation

The graph of fatigue crack length versus number of cycles usually has the form shown in figure 5.1

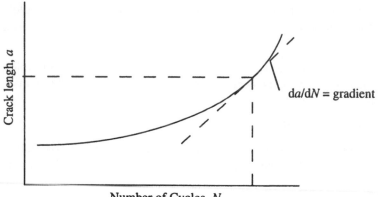

Figure 5.1

The graph may be used to calculate how fast a crack is growing at any particular point by measuring the gradient of the curve at the point of interest. This measure is the crack-growth-increment per cycle, da/dN, and its derivation is demonstrated in figure 5.1.

For quasi-elastic (LEFM) conditions, the most popular, empirical relationship between the crack-growth-increment per cycle (da/dN) and parameters of stress range ($\Delta\sigma$) and the instantaneous crack length (a) is that proposed initially by Paris and Erdogan which gives:

$$da/dN = A\Delta K^m \tag{54}$$

where A and m (the Paris exponent) are constants, and ΔK is the range of stress-intensity factor:

$$\Delta K = K_{max} - K_{min} \tag{55}$$

Here, K_{max} and K_{min} are respectively the maximum and minimum values of stress intensity factor in the cycle, derived from the maximum or minimum stress and the appropriate compliance function, $Y(a/W)$ (see section 3.2), for an instantaneous crack length, a.

The engineering application of Equation 5.4 may be illustrated by considering the case of an infinite body, containing a central, through thickness crack of length $2a$ lying normal to a constant stress-range, $\Delta\sigma$ (cf. section 2.3). The expression for ΔK then becomes:

$$\Delta K = \Delta\sigma(\pi a)^{1/2} \tag{56}$$

(see table 2.1) so that equation 54 may be written as:

$$da/dN = A\Delta\sigma^m \pi^{m/2} a^{m/2} \tag{57}$$

Variables (a and N for constant $\Delta\sigma$) may be separated to give:

$$a^{-m/2}da = A\Delta\sigma^m \pi^{m/2} dN \tag{58}$$

It is then possible to integrate the two sides of this equation: the left hand side between limits a_0 (the initial defect size, determined by process control or NDT inspection limit) and a_f (the final defect size, determined by the onset of fast fracture or plastic collapse): the right hand side between limits zero and the number of cycles to failure, N_f. The result of this integration, for all values of m except $m = 2$, is:

$$\frac{2}{(2-m)}\left| a_f^{1-m/2} - a_o^{1-m/2}\right| = A\,\Delta\sigma^m \pi^{m/2} N_f \tag{59}$$

For m = 2, the result is

$$\ln\left(a_f/a_0\right) = A\Delta\sigma^2\,\pi\,N_f \tag{60}$$

Equivalent expressions for finite geometries may be obtained using the compliance functions, $Y(a/W)$.

The initial defect size is determined by the NDT detection limit or the size of defect introduced by the processing of the material. The final defect size can be calculated from the fracture toughness of the material. Hence, for a given applied stress range, $\Delta\sigma$, equations 59 and 60 can be used to calculate the propagation lifetime, N_f. Increases in N_f are expected if a_f can be increased (e.g. by using material of higher fracture toughness) or if a_0 is reduced (e.g. by a process which reduces the size of initial defects). For higher values of m, N_f is particularly sensitive to the value of a_0, which may initially be taken as a constant for a given material, or structure.

Example 5a - Effect of Crack Size on Fatigue Lifetime

An NDT evaluation of a component shows that there are cracks present of 3 mm which grew by fatigue to a final crack length under service conditions of 8 mm, determined by fast fracture. Calculate the percentage increase in fatigue life for the component

(i) If the final crack length is extended (by 2 mm to 10 mm) through the use of a material of higher fracture toughness.

(ii) If the initial crack size is reduced (by 2 mm) to 1 mm.

Take the Paris exponent, *m*, to be 3 for the material in question.

Solution

From equation 59

$$\frac{2}{(2-m)}\left|a_f^{1-m/2} - a_o^{1-m/2}\right| = A\,\Delta\sigma^m \pi^{m/2} N_f$$

If $m = 3$

$$-2\left|a_f^{-1/2} - a_o^{-1/2}\right| = A\Delta\sigma^3 \pi^{3/2} N_f$$

For $a_0 = 3$ mm and $a_f = 8$ mm

$$A\,\Delta\sigma^3\, N_f = 2.54$$

(i) For $a_0 = 3$ mm and $a_f = 10$ mm

$$A\,\Delta\sigma^3\, N_f = 2.97$$

$$\frac{N_f(3-10\text{mm})}{N_f(3-8\text{mm})} = \frac{2.97}{2.54} = 1.17$$

This gives an increase in fatigue life of 17%.

(ii) For $a_0 = 1$ mm and $a_f = 8$mm

$$A\,\Delta\sigma^3\, N_f = 7.34$$

$$\frac{N_f(1-8\text{mm})}{N_f(3-8\text{mm})} = \frac{7.34}{2.54} = 2.89$$

This gives an increase in fatigue life of 189%.

Comment

It can be seen that the life of the component is very much more sensitive to the value of initial crack length. If the production process can be improved to reduce the initial defect size the component can be left in service for nearly three times the original lifetime. An improvement in material toughness (of the order of 12%) can be seen to have only a small effect (17%). Part (i) of the calculation shows that 17% of the life is taken up growing from 0.8

a_f to a_f (8 mm to 10 mm). Quite large safety factors are applied to determine 'safe lifetimes' for structures and for service assessments, 17% is hardly significant.

From the example above, we can see that the initial defect size has a very marked effect on the fatigue lifetime of the component. In all materials the defects will have a spread of sizes and it is the maximum defect size which determines the life of a batch of components. When a component is guaranteed to last for a particular lifetime it is the maximum defect size which gives the minimum lifetime. Thus, a reduction in the average defect size in a material will only have an influence on the predicted minimum lifetime if the maximum defect size in high stress regions is reduced.

5.2 The da/dN Curve

From equation 54 we may take logarithms of both sides to obtain
$\log (da/dN) = m \log (\Delta K) + \text{const.}$
Hence the graph of $\log (da/dN)$ against $\log \Delta K$ is predicted to be a straight line. The general behaviour of materials can be seen in figure 5.2. Although the Paris relation (equation 54) can be seen to be obeyed in the central portion of the graph, the ends deviate from linear. At low ΔK it can be seen that there is a minimum value of ΔK below which no fatigue crack growth occurs. This is called the *threshold* value. At high ΔK, the crack growth accelerates to failure at a faster rate than is predicted by the Paris relation.

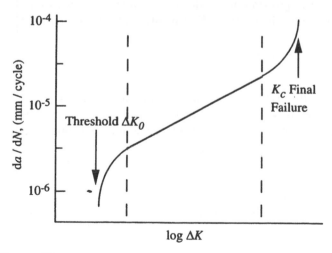

Figure 5.2

The Paris exponent (the gradient of the central portion of the graph) varies from material to material. General values can be found in table 5.1.

Material	Paris Exponent (m)	Constant (A)
Average "Steel"	3	10^{-11}
Structural steel (A533B, wet)	3	4×10^{-11}
Forging steel	2 - 3	10^{-11}
High strength Al alloy (7010)	3	10^{-12}
Ni alloy	3.3	4×10^{-12}
Ti alloy (IMI 834)	5	10^{-11}

Table 5.1. Table of Paris exponents for various alloys.

Example 5b - Calculation of the Fatigue Lifetime of an Aero-Engine Disc

An aero-engine gas-turbine disc contains defects which may be treated as internal cracks of half length, a = 50 µm. The critical crack half length in the material is 2 mm. During each flight, of average duration 3 hours, the disc is subjected to one large 'take-off / landing' cycle, which produces a stress amplitude of 1000 MPa. The crack growth rate, da/dN, (in m/cycle) is related to the alternating stress intensity, ΔK, (in MPa m$^{1/2}$) through the expression:

$$da/dN = 4 \times 10^{-12} (\Delta K)^3$$

Calculate:
(i) the life of the disc limited by the take-off landing cycles;
(ii) the maximum value of the vibrational stress which could be allowed in an engine running at 5000 revolutions per minute, if the life of the disc is not to be reduced by more than 5%.

Solution
Using MPa and m.
(i) If we assume that the crack is treated as a crack in an infinite plate then

$$\Delta K = \Delta\sigma(\pi a)^{1/2} \qquad \text{(see table 2.1)}$$

and the integration of $da/dN = 4 \times 10^{-12} (\Delta K)^3$

becomes $\quad \dfrac{2}{(2-3)}\left|a_f^{-1/2} - a_o^{-1/2}\right| = 4x10^{-12} (1000)^3 \pi^{3/2} N_f$

(see equation 59).

This gives $\qquad -2|22.36 - 141.42| = 4x10^{-3} \pi^{3/2} N_f$

and $\qquad\qquad N_f = 10690$ **cycles**.

(ii) If the life span can be reduced by a maximum of 5%, then this is 534 cycles.

For an average flight time of 3 hours and a vibrational frequency of 5000 rpm, the number of cycles is then:

$$534 \times 3 \times 60 \times 5000 = 4.806 \times 10^8 \text{ cycles.}$$

Now $\qquad \dfrac{2}{(2-3)}\left|a_f^{-1/2} - a_o^{-1/2}\right| = 4x10^{-12} (1000)^3 \pi^{3/2} N_f$

As the defect sizes need to be calculated from the large stress cycle data, the equation becomes:

$$\dfrac{2}{(2-3)}\left|a_f^{-1/2} - a_o^{-1/2}\right| = 4x10^{-12} (1000)^3 \pi^{3/2} 534$$

For the vibrational stress we have:

$$\dfrac{2}{(2-3)}\left|a_f^{-1/2} \quad a_o^{-1/2}\right| = 4x10^{-12} (\Delta\sigma)^3 \pi^{3/2} 4.806x10^8$$

As the left hand sides of the equations are equal for the vibrational stress causing 5% crack growth of the take-off / landing cycle, then

$$4x10^{-12} (\Delta\sigma)^3 \pi^{3/2} 4.806x10^8 = 4x10^{-12} (1000)^3 \pi^{3/2} 534$$

$$(\Delta\sigma)^3 = (1000)^3 \, 534 \, / \, 4.806x10^8$$

$$(\Delta\sigma)^3 = 1111.1$$

$$\Delta\sigma = \textbf{10.4 MPa}$$

The maximum vibrational stress amplitude which would reduce the life of the turbine disc by the specified amount is 10.4 MPa.

5.3 Application to Variable Amplitude Loading

As we have seen in Example 5a, in pragmatic engineering assessment of fatigue life, a_f may be taken as approximately constant, because the number of cycles associated with growth from, say, 0.8 a_f to a_f is small compared with N_f, and the calculation of lifetime is therefore insensitive to the precise value taken for a_f. It is possible to provide a good engineering approximation to the calculation of lifetime under variable amplitude loading. It is necessary to reduce the general fatigue spectrum to a number of 'blocks' of constant amplitude. If a given block is denoted by i, we count the number of cycles in that block as N^i and the number of cycles to failure at the appropriate stress amplitude as N^i_f. There is a traditional (non fracture mechanics) method used to calculate the failure criterion under these conditions (Miner's Law) and it may be shown that the integration crack growth rates leads to a compatible result.

If equation 58 is integrated between crack lengths of a_0 and a_1, for a fixed stress range, the result can be denoted as I_{01}. If the crack grows to failure at a_f then the integral becomes I_{0f}. This enables us to write

$$\frac{N^1}{N^1_f} = \frac{I_{01}}{I_{0f}} \tag{61}$$

If the stress range is changed and the crack grows from a_1 to a_2 the integration now gives I_{12} and

$$\frac{N2}{N^2_f} = \frac{I_{12}}{I_{0f}} \tag{62}$$

In general

$$\sum \frac{N^i}{N^i_f} = \frac{1}{I_{0f}} \left(I_{01} + I_{12} + I_{23} + \ldots + I_{f-1,f} \right) \tag{63}$$

The term in parentheses on the right hand side of equation 63 are simply the sequential integrals of $a^{-m/2}$ from a_0 to a_1 to a_2 to a_3 . . . etc. to a_f. By the definition of integrals their sum is equal to I_{0f}, so that the resultant right hand side is unity. Hence:

$$\sum \frac{N^i}{N^i_f} = 1 \tag{64}$$

This is known as Miner's Law.

Example 5c The Use of Miner's Law

Use Miner's law to calculate the minimum stress allowed in part ii of example 5b

Solution

As the lifetime must not be reduced by more than 5% the value of N^i for the high stress cycle becomes at least $0.95 \, N^i_f$. If the number of cycles at the lower stress level are taken as n^i and n^i_f, then Miner's law states that,

$$\frac{0.95N_f^i}{N_f^i} + \frac{n^i}{n_f^i} = 1$$

Hence, $\qquad n^i = 0.05n_f^i$.

As $N_f^i = 10690$ cycles, and the average flight time is 3 hours and the vibrational frequency is 5000 rpm, then the number of vibrational cycles, n^i is

$(10690 \times 0.05) \times 3 \times 60 \times 5000 = 4.806 \times 10^8$ cycles.

Using the equations

$$da / dN = 4 \times 10^{-12} (\Delta K)^3$$

and $\qquad \Delta K = \Delta\sigma(\pi a)^{1/2}$

we obtain $\qquad \dfrac{2}{(2-3)}\left|a_f^{-1/2} - a_o^{-1/2}\right| = 4x10^{-12} (\Delta\sigma)^3 \pi^{3/2} n_f^i$

Substituting for the final and initial defect sizes (see Example 5b) and n_f^i we obtain

$$-2|22.36 - 141.42| = 4x10^{-12} (\Delta\sigma)^3 \pi^{3/2} \frac{4.806x10^8}{0.05}$$

Hence $\qquad\qquad \Delta\sigma^3 = 1112.2$

and $\qquad\qquad\quad \Delta\sigma = 10.4 \text{ MPa}$

Comment
The use of Miner's law predicts the same maximum vibrational stress amplitude as that in Example 5b.

5.4 Application to Quality Control of Materials and / or Fabrication Procedures

Fatigue life calculations may be used to calculate the maximum permissible flaw size in a component to prevent failure in a certain number of cycles for a particular stress range. This can determine either the time a component may be left in service *or* the period between NDT inspections.

Example 5d - Calculation of Minimum Defect Size

The aero-engine-gas turbine disc in Example 5b is now made of a different material which can now be loaded up to a stress amplitude of 1200 MPa. If the crack growth equation remains the same, what is the control of the initial defect size which would ensure the same component lifetime?

Solution

Taking the equations $\quad\quad da/dN = 4 \times 10^{-12} (\Delta K)^3$

and $\quad \dfrac{2}{(2-3)} \left| a_f^{-1/2} - a_o^{-1/2} \right| = 4 \times 10^{-12} (1200)^3 \pi^{3/2} N_f$

from Example 5b,

we have $\quad -2 \left| 23.36 - a_0^{-1/2} \right| = 6.912 \times 10^{-3} \pi^{3/2} 10690$

$$a_0^{-1/2} = \frac{412.5}{2} + 23.36 = 229.6$$

$$a_0 = 19 \ \mu m$$

This is the half length of the embedded crack, so the full length is 38 μm.

Comments

Such sub-surface defects are virtually impossible to detect using NDT techniques. Thus, the defect size must be controlled using process control. This involves clean melts, atomised powder and a fine sieve size, with guaranteed consolidation to remove porosity. This defect size at the lower stress level would give a life of 18592 cycles.

In Example 5b we showed that the change from a 1 mm to a 3 mm initial crack size has a marked effect on the fatigue life. This may be generalised to show that a distribution of initial defect sizes, a_0 causes a distribution in the number of cycles to failure. Not only does the fatigue life depend on the initial defect size, it also depends on the orientation of the defect to the direction of the applied stress and on the likelihood of finding a defect of a particular size within the vicinity of a stress concentrator.

All these features may be present in welded joints, in which there may be crack-like defects within the weld, residual stresses and stress concentrators. Situations of this kind can be approached by a weight function method in which the various stress concentrators can be represented and added to the functions for the residual stress within the weld and that of a sharp crack at the tip of the stress concentrator.

5.5 Monitoring of Fatigue Crack Length

Fatigue crack growth usually takes place over a large number of cycles, and the crack growth in a testpiece needs to be measured if a graph of crack growth per cycle versus stress intensity is to be plotted. This is done in a number of ways. One method is direct optical observation of the crack in which the side of the testpieces are polished to enable easier observation. The position of the crack tip can then be monitored using a microscope. Another method, which requires the test to be interrupted, is the replication technique. A sheet of acetate, wetted with acetone, is placed on the polished side of the testpiece and allowed to harden. Once removed, the sheet can be viewed under a microscope and the crack can be seen in relief.

For metals a popular method of crack monitoring is the potential drop technique. An a.c. or d.c. current is applied to the testpiece and the potential drop across the crack is monitored on the top surface either side of the notch. The voltage is divided by the voltage at a crack length of 0.25 W for the SEN bend testpiece and 0.35W for the CTS testpiece. The crack length during fatigue can be calculated from the calibration curves shown in figure 5.3. With all these methods the instantaneous crack growth must be averaged over a number of points to remove any contribution from noise.

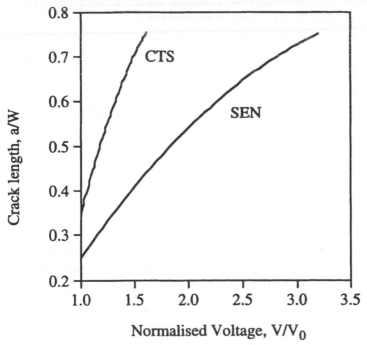

Figure 5.3 Voltage Calibrations for the CTS and SEN Testpieces (ref. 22)

5.6 Effect of Residual Stresses

Residual stresses may be present in a component because of the manufacturing route employed. These stresses may not be detrimental as they can often oppose the in-service stresses on the component, thereby prolonging the life of the component. Beneficial compressive residual stresses may be deliberately introduced into components by methods such as shot-peening and autofrettaging.

Example 5e - Effect of Residual Stress

A component is manufactured from a steel in which the fatigue crack growth is characterised by the equation

$$da/dN = 10^{-11}(\Delta K)^3$$

For the component in question the stress intensity is given by $K = 1.12\, \sigma_{app}\, (\pi a)^{1/2}$ (for an edge crack), the initial defect size is 1 mm and final failure occurs when the crack length is 10 mm. If the in service stress has a range of 400 MPa and a peak stress of 700 MPa, what is the increase in fatigue life if a uniform residual stress of 500 MPa opposes the crack opening?

Solution

From
$$da/dN = 10^{-11}(\Delta K)^3$$

and
$$K = 1.12\, \sigma_{app}\, (\pi\ a)^{1/2}$$

we have
$$\frac{2}{(2-3)}\left| a_f^{-1/2} - a_o^{-1/2} \right| = 10^{-11}\, 1.12^3\, (\sigma)^3\, \pi^{3/2} N_f$$

Then
$$N_f = \frac{5.528 \times 10^{11}}{\Delta\sigma^3}$$

In the untreated case $\Delta\sigma$ is 400 MPa and N_f is **8637 cycles.**

The residual stress will close the crack until a stress of 500 MPa is reached. This reduces the stress amplitude to 200 MPa and hence the number of cycles to failure becomes **69100**. This is an eight fold increase in fatigue life.

The residual stress distribution within a component is not necessarily uniform, and hence the calculation of the fatigue lifetimes become more difficult. Weight functions (see section 2.7) are widely used for this type of calculation.

Example 5f - The effect of residual stress on fatigue crack growth in an autofrettaged gun barrel.

A gun barrel with an outside diameter of 60 mm and a bore of 30 mm is made of a NiCrMoV forging steel. When the gun is fired a pressure of 345 MPa is developed within the barrel and craze cracks due to the heating and rapid cooling of the inner surface of barrel form and propagate with each successive firing. If the material has a minimum fracture toughness of 75 MPa m$^{1/2}$, calculate

(i) how many times the gun can be fired if the initial defect size is 3 mm and the fatigue crack growth is represented by the equation

$$\frac{da}{dN} = 3 \times 10^{-11} \Delta K^3$$

(ii) the crack growth rate at 3 mm.

If the barrel is autofrettaged (the bore is cold-expanded using an oversize mandrel so that a substantial part of the wall thickness is yielded plastically) calculate the crack growth rate at 3 mm in this case. The residual stress intensity is given by

$$K_{res} = f\,\sigma_{YS}\,\sqrt{\pi a}$$

where f is the weight function and is given as -0.642 for a radial crack in a cylinder. The yield strength is 1050 MPa.

Solution

(i) The stress intensity for small radial cracks in a thick-walled cylinder is given by

$$K = 1.12\,P\sqrt{\pi a}\,\frac{2R_o^{\,2}}{R_o^{\,2} - R_i^{\,2}}$$

where P is the pressure. For $R_O = 2R_i$ this becomes

$$K = 3P\sqrt{\pi a}$$

As the cracks will be semi-circular rather than through-thickness we use the factor 0.64 and hence

$$K = 2P\sqrt{\pi a}$$

Rearranging to calculate a, the crack length at failure, we have

$$a = \frac{1}{\pi}\left(\frac{K_{Ic}}{2P}\right)^2$$

which gives a value of a of 3.76 mm.

The equation for fatigue crack growth is

$$\frac{da}{dN} = 3\times10^{-11}\Delta K^3$$

As
$$K = 2P\sqrt{\pi a}$$

we obtain
$$-2\left[\frac{1}{\sqrt{a_f}} - \frac{1}{\sqrt{a_0}}\right] = 24\times10^{-11}\, P^3\, \pi^{3/2}\, N_f\, .$$

$$N_f = -2\left[\frac{1}{\sqrt{0.00376}} - \frac{1}{\sqrt{0.003}}\right]\frac{1}{24\times10^{-11}\,345^3\,\pi^{3/2}}$$

With an initial defect size of 3 mm and a final defect size of 3.76 mm we obtain a value for N_f of 71 cycles. The gun can be fired 71 times before it breaks.

(ii) The fatigue crack growth rate is given by

$$\frac{da}{dN} = 3\times10^{-11}\Delta K^3$$

which can be written as

$$\frac{da}{dN} = 3\times10^{-11}\left(2P\sqrt{\pi a}\right)^3$$

For a crack length of 3 mm, this becomes

$$\frac{da}{dN} = 3\times10^{-11}\left(2\times345\sqrt{\pi 0.003}\right)^3$$

$$\frac{da}{dN} = 3\times10^{-11}(66.99)^3$$

$$\frac{da}{dN} = 9.0\times10^{-6}$$
$$\text{m/cycle}$$

The crack grows at **9μm / cycle.**

If the gun is autofrettaged then the residual stress intensity is given by

$$K_{res} = f\,\sigma_{YS}\,\sqrt{\pi a}$$

At 3 mm, with a yield strength of 1050 MPa and a weight function of -0.642 we obtain

$$K_{res} = -0.642\times1050\sqrt{\pi 0.003}$$

$$K_{res} = -65.4.$$

This is for through-thickness cracks, so we have to multiply by the factor for semi-circular cracks (0.64) to obtain a residual stress intensity of -41.9 MPam$^{1/2}$.

We can now add the residual stress intensity to the applied stress intensity to calculate the modified crack growth rate.

$$\frac{da}{dN} = 3 \times 10^{-11} \left(\left(2P\sqrt{\pi a}\right) - 41.9 \right)^3$$

$$\frac{da}{dN} = 3 \times 10^{-11} (66.99 - 41.9)^3$$

$$\frac{da}{dN} = 5 \times 10^{-7} \, \text{m/cycle}$$

The modified crack growth rate is 0.5µm/cycle. This is nearly 20 times lower than for the non-autofrettaged case.

Comments

In the untreated case the barrel will split on firing when the crack attains a length of 3.74 mm. However, the life of the gun has been greatly enhanced by the autofrettage process as can be seen by the relevant crack growth data. It has been shown experimentally that the number of cycles to failure of an autofrettaged gun barrel can be more than doubled.

6. Concluding Remarks

An attempt has been made to introduce the reader to the subject of Fracture Mechanics through selected Worked Examples. Much background theory has been omitted; the aim being to show how fracture toughness values are derived in practice and how these values are utilised in defect assessment and design.

Linear Elastic Fracture Mechanics may be applied with confidence in situations where failure occurs under essentially elastic conditions. The critical value of stress intensity, K_{IC}, is used in calculating either the permissible defect size for a given operating stress or, alternatively, the maximum applied stress for a given defect size.

When significant localised plastic deformation accompanies crack extension, the LEFM approach is invalid, and recourse can be made to crack-opening-displacement methods, to J-integral methods, or to other mixed elastic / plastic characterisations. Examples have been given to demonstrate the Welding Institute and (former) Central Electricity Generating Board approaches to defect assessment in situations where extensive plasticity occurs.

Where the failure event does not result from a fracture process, localised at the crack tip, but is one of gross plastic instability across the whole section, fracture toughness parameters such as K_{IC} or δ_c, are not relevant design criteria. Under these conditions the onset of failure is governed by the stiffness of the loading configuration, the material's flow stress, and geometrical parameters, predominantly the area ahead of the crack, allowing for any slow crack growth. It should be emphasised that a value of δ_m, or the load at plastic instability obtained on a small, ductile testpiece, may have no relevance whatsoever to the failure conditions obtained in a large structure made of the same material, and any 'toughness' values based on these parameters should be treated with extreme caution.

The conditions imposed in the laboratory, where loads, testpiece dimensions and compliance factors are able to be determined with high accuracy, are such that errors in the toughness are small, and the ±10% scatter observed in experimental values of K_{IC} is due primarily to material variability. Use of these values in the practical assessment of defects in structures is, however, often subject to larger errors, because neither the 'applied stress' nor the 'defect size' in service is able to be calculated with absolute precision.

The nominal applied stress, e.g. the hoop stress in a pressure vessel shell, is, of course, calculated accurately, and is the basis of the engineer's design against gross plastic collapse, but the 'applied stress' in the regions of importance with respect to fast fracture, i.e. around stress-concentrators, which usually involve welded details and residual stresses, is far less well specified. Few elastic / plastic, plane strain, stress analyses are available for such features at present, although the increasing use of finite-element methods should produce more detailed information in the near future.

The specification of 'defect size' is no simple matter, especially when several, closely-spaced indications are obtained, because interaction effects between defects are not fully understood. Present practice equates the effect of a cluster of defects to that of a sharp crack whose profile is the ellipse or semi-ellipse that circumscribes the defects. The detection of defects is particularly difficult in clad vessels, in welded nozzle regions, or in areas where access is difficult. For example, 'nodes' of oil rigs, where several tubes are welded together, represent a very awkward geometrical configuration. Another problem, which arises with ultra-high-strength materials, is that the calculated critical defect size may be smaller than the resolution limits of conventional non-destructive testing techniques.

Nevertheless, despite these limitations, Fracture Mechanics is proving to be a powerful and important tool for both defect assessment and the fracture-safe design of components and structures. This is particularly true for situations where unscheduled plant breakdown cannot be tolerated, or where the potential hazard of a failure is such that the highest possible standards of structural integrity are required. Such areas include electrical power generation, by fossil-fuel or nuclear means; military projects, where high design stresses are used to maximise performance; aircraft and transportation; and safety-critical components, such as lifting-gear, hydraulic pit-props, and other pressurised cylinders, such as those used for gas storage. The application of Fracture Mechanics principles to lower strength structural steels in general engineering fields, such as pipe-lines, pressure-vessels, bridges and ships, is also being actively pursued.

With the establishment of official bodies such as the Health and Safety Executive, and the Nuclear Installations Inspectorate, and the rapid growth, throughout industry, of Quality Assurance procedures, there is no doubt that positive steps are being taken to avoid unexpected failures in service. The tragic consequences of a large failure, such as the Flixborough disaster, are only too obvious. Although the critical catastrophic event in that particular instance was not attributed to fast fracture, it is clear that, in general, the techniques of Fracture Mechanics are vitally important in the assessment of a structure's integrity.

At the very least, engineers now have to recognise that their structures, far from being the perfect solids assumed in traditional design textbooks, contain a distribution of crack-like defects, but can, nevertheless, be operated in a perfectly safe manner, provided that critical crack-growth conditions are not exceeded. The metallurgist must recognise that these 'new-fangled' toughness parameters which are being specified do have real meaning in the establishment of a quantitative design against fracture and must strive to understand the ways in which toughness may be altered by metallurgical treatments, in order to maximise the performance of a material in service.

7. References

1   J.F. KNOTT: *Fundamentals of Fracture Mechanics*, Butterworth (London), 1973, third impression 1979. Provides a general background to the principles of fracture mechanics.

2.   *Fracture Toughness Testing and its Applications*, STP 381 ASTM, Philadelphia, PA, 1965.

3.   *Fracture Toughness*, STP 514 ASTM, Philadelphia, PA, 1971.

4.   *Journal of Strain Analysis* **10**, *4* 1975 gives further details.

5.   D.D. ROOKE and D.J. CARTWRIGHT: *Compendium of Stress Intensity Factors,* HMSO, London, 1975, contains values of K for a large number of configurations

6.   BS 7448:1991, *Methods for determination of K_{Ic}, critical CTOD and critical J values of metallic materials*, BSI Standards, London.

7.   PD 6493:1991, *Guidance on methods for assessing the acceptability of flaws in fusion welded structures.* BSI Standards, London.

8.   R.W. NICHOLS *et al.*, eds. *Practical Fracture Mechanics for Structural Steels*, UKAEA/Chapman and Hall, London, 1969.

9.   D. ELLIOTT, E.F. WALKER, R.R. BARR and P. TERRY: 'Measurement of COD and its Application to Defect Assessment,' *Metal Construction*, December 1976.

10.   W.J. JACKSON and J.C. WRIGHT: *Metals Technology*, September 1977, p. 425.

Practical Application of Fracture Mechanics to Pressure Vessel Technology, Institute of Mechanical Engineers, 1971, papers by:

11.   F.M. BURDEKIN and M.G. DAWES: p. 28.

12.   R.F. SMITH and J.F. KNOTT: p. 65.

13.   D. ELLIOTT, E.F. WALKER and M.J. MAY: p. 217.

Tolerance of Flaws in Pressurised Components, Institute of Mechanical Engineers, 1978, papers by:

14.   M.G. DAWES and M.S. KAMATH: p. 27.

15.   G. CLARK, S.M. ELSOUDANI, W.G. FERGUSON, R.F. SMITH and J.F. KNOTT: p. 121.

16.   T. INGHAM and J. SUMPTER, p. 161.

17.   I. MILNE, K. LOOSEMORE and R.P. HARRISON: p. 317.

18. I. MILNE, R.A. AINSWORTH, A.R. DOWLING and A.T. STEWART: *Assesment of the integrity of structures containing defects*, Central Electricity Generating Board, 1976 (R/H/R6-Rev3).
The Welding Institute approach (Section 4.7, Example 4i) may be followed in references 11 and 14. The CEGB approach (Section 4.8, Example 4j) is described in reference 17.
19 I.D. LEWIS, R.F. SMITH and J.F. KNOTT: *International Journal of Fracture* **11** 1975, pp 179-183.
20 X.-R. WU and A.J. CARLSSON: *Weight Functions and Stress Intensity Factor Solutions,* Pergamon Press (Oxford), 1991.
21 J.E. SRAWLEY and B. GROSS: Side Cracked Plates Subject to Combined Direct and Bending Forces: *Cracks and Fracture*, STP 601, ASTM, 1976, pp 559-579.
22 R.O. RITCHIE and K.J. BATHE: *International Journal of Fracture* **15** 1979, pp 47-55.

Appendix - Derivation of the Displacement Within a Crack

If a plate contains a crack (see figure 2.1) then the stress across the x-axis in Figure 2.3 has to have a distribution such that the boundary conditions are met. These boundary conditions are:
(i) at large distances from the crack the stress must approach the applied stress, σ_{app},
(ii) close to the crack tip the stress must be very high,
(iii) within the crack the stress must be zero.
A simple form of expression to satisfy the first two requirements would be

$$\sigma = \frac{\sigma_{app}}{1 - a/x} \tag{A1}$$

or for symmetry

$$\sigma = \frac{\sigma_{app}}{\sqrt{1 - a^2/x^2}} \tag{A2}$$

Although this is identical to Equation 3 (Section 2.3) it does not meet the third requirement. To enable the equation to be analytic over the whole range of x, a complex number, $c = x + iy$, replaces x and then only the real part of the arising complex function is taken:

$$\sigma = \frac{\operatorname{Re}\sigma_{app}}{\sqrt{1 - a^2/c^2}} = \operatorname{Re}\phi \tag{A3}$$

where

$$\phi = \frac{\sigma_{app}}{\sqrt{1 - a^2/c^2}} \tag{A4}$$

However, because a rigorous derivation for the displacement within a crack is required for Example 2d, the equation has to be written

$$\sigma_2 = \frac{\operatorname{Re}\sigma_{app}}{\sqrt{1 - a^2/c^2}} = \operatorname{Re}\phi \tag{A5}$$

Where σ_2 is the stress in the direction perpendicular to the crack plane (or parallel to σ_{app}) and x_1 is the direction of crack growth. x_2 and x_3 are the other orthogonal axes.

The above equation holds for all values of $|x_1| > |a|$.

To calculate the virtual work done in extending a crack a small amount it is necessary to calculate the displacement in the x_2 direction, u_2. This is related to the strain, ε_2, by

$$\varepsilon_2 = \frac{du_2}{dx_2} \tag{A6}$$

and the strain is related to the stress by the following equations

$$E\,\varepsilon_1 = \sigma_1 - \nu(\sigma_2 + \sigma_3), \quad E\,\varepsilon_2 = \sigma_2 - \nu(\sigma_1 + \sigma_3) \text{ and } E\,\varepsilon_3 = \sigma_3 - \nu(\sigma_2 + \sigma_1) \tag{A7}$$

where ν is Poisson's ratio and E is Young's modulus. In plane strain $\varepsilon_3 = 0$, then

$$E\,\varepsilon_2 = (1 - \nu^2)\,\sigma_2 - \nu\,(1 + \nu)\,\sigma_1. \tag{A8}$$

Although there is an equation for σ_2 already derived it cannot be directly substituted into the equation above as the displacement is in the region $|x_1| < |a|$ and thus involves the imaginary part of ϕ. To obtain the complete analytic functions for σ_1 and σ_2 we first employ the Airy stress equations:

$$\sigma_1 = \frac{d^2\Phi}{dx_2{}^2}, \quad \sigma_2 = \frac{d^2\Phi}{dx_1{}^2} \quad \text{and} \quad \sigma_{12} = -\frac{d^2\Phi}{dx_1 dx_2}, \tag{A9}$$

where Φ is the Airy stress function, and is a function of ϕ which satisfies the equation:

$$\nabla^4\Phi = 0 \tag{A10}$$

where

$$\nabla^4\Phi = \frac{\delta^4\Phi}{\delta x_1{}^4} + 2\frac{\delta^4\Phi}{\delta x_1{}^2 \delta x_2{}^2} + \frac{\delta^4\Phi}{\delta x_2{}^4} \tag{A11}$$

The form of Φ chosen is

$$\Phi = \int\int \operatorname{Re}\phi\, dx_1 dx_1 + x_2 \int \operatorname{Im}\phi\, dx_1 \tag{A12}$$

The Airy stress equations can be used to obtain values of σ_1 and σ_2. It is straightforward to obtain a value for σ_2 as the above equation only needs to be differentiated twice with respect to x_1. Hence

$$\sigma_2 = \mathrm{Re}\,\phi + x_2 \frac{d(\mathrm{Im}\,\phi)}{dx_1} \qquad (A13)$$

To obtain σ_1, Φ must be differentiated with respect to x_2 and the Cauchy-Riemann equations have to be employed. These are

$$\frac{d(\mathrm{Re})}{dx_1} = \frac{d(\mathrm{Im})}{dx_2}, \quad \frac{d(\mathrm{Im})}{dx_1} = -\frac{d(\mathrm{Re})}{dx_2}, \qquad (A14)$$

and give

$$\sigma_1 = \mathrm{Re}\,\phi - x_2 \frac{d(\mathrm{Im}\,\phi)}{dx_1} \qquad (A15)$$

The **first term** in the equation for Φ is chosen to match the form $\sigma_2 = \mathbf{Re}\ \phi$ as derived earlier and the *second term follows to ensure that equation A10 is satisfied*. The Airy stress equations (A12, A14) are then substituted into the equation for ε_2:

$$E\varepsilon_2 = (1+v)\left[(1-v)\left(\mathrm{Re}\,\phi - x_2 \frac{d(\mathrm{Im}\,\phi)}{dx_1}\right) - v\left(\mathrm{Re}\,\phi + x_2 \frac{d(\mathrm{Im}\,\phi)}{dx_1}\right)\right] \qquad (A16)$$

and from $\varepsilon_2 = du_2 / dx_2$ we have

$$Eu_2 = (1+v)\int\left[(1-v)\left(\mathrm{Re}\,\phi - x_2 \frac{d(\mathrm{Im}\,\phi)}{dx_1}\right) - v\left(\mathrm{Re}\,\phi + x_2 \frac{d(\mathrm{Im}\,\phi)}{dx_1}\right)\right]dx_2 . \quad (A17)$$

Using the Cauchy-Riemann equations (A14) it follows that

$$E\,u_2 = (1 + v)\,[2(1 - v)\int \mathrm{Im}\ \phi\ dx_1\ - x_2\ \mathrm{Re}\ \phi]. \qquad (A18)$$

In this case $x_2 = 0$, hence:

$$u_2 = 2(1 - v^2)\int \mathrm{Im}\ \phi\ dx_1 / E \qquad (A19)$$

As

$$\mathrm{Im}\,\phi = \frac{i\sigma_{app}}{\sqrt{1 - a^2/x_1^2}} \qquad (A20)$$

then

$$\int \mathrm{Im}\ \phi\ dx_1 = i\,\sigma_{app}\,\sqrt{(x_1^2 - a^2)} = \sigma_{app}\,\sqrt{(a^2 - x_1^2)} \qquad (A21)$$

and u_2 becomes

$$u_2 = 2(1 - v^2)\,\sigma_{app}\,\sqrt{(a^2 - x_1^2)} / E. \qquad (A22)$$

By changing the origin to the crack tip, by making $x_1 = r + a$, using the equation $K = \sigma_{app} \sqrt{(\pi a)}$ and defining α as $(1 - v^2)$, u_2 becomes:

$$u_2 = \frac{2K\alpha\sqrt{\dfrac{a^2 - (r+a)^2}{\pi a}}}{E} \qquad (A23)$$

Now if the crack is advanced to $a + \delta a$ the equation becomes

$$u_2 = \frac{2K\alpha\sqrt{\dfrac{(a+\delta a)^2 - (r+a)^2}{\pi a}}}{E} \qquad (A24)$$

As r is small $r^2 + 2ra \rightarrow 2ra$ and $\delta a^2 \rightarrow 0$ then this simplifies to

$$u_2 = 2\sqrt{\frac{2}{\pi}}\alpha\frac{K}{E}\sqrt{\delta a - r} \qquad (A25)$$

which is the equation used in Example 2d - The Virtual Work Theorem.